TH
RI
FOR YOU

THE RIGHT DOG FOR YOU

David Alderton
with breed photography by Marc Henrie

HOW TO CHOOSE THE PERFECT BREED FOR YOU AND YOUR FAMILY

First published in the United Kingdom
in 2021 by
Ivy Press
An imprint of The Quarto Group
The Old Brewery, 6 Blundell Street
London N7 9BH, United Kingdom

ISBN: 978-0-7112-5750-4
Ebook ISBN: 978-0-7112-5751-1

This book was conceived,
designed, and produced by
Ivy Press
58 West Street, Brighton BN1 2RA,
United Kingdom

Publisher Richard Green
Editorial Director Jennifer Barr
Art Director Paileen Currie
Project Editor Joe Hallsworth
Design JC Lanaway
Picture Research Bella Skertchly

Printed in Singapore
10 9 8 7 6 5 4 3 2 1

CONTENTS

INTRODUCTION

Just a glance at the line-up like the one below highlights the extraordinary diversity that exists within dog breeds. Dogs are universally possessed of boundless character and sharp intelligence, but every breed and every individual dog has its own unique appearance and traits, since they come in all manner of shapes, colours, patterns and sizes, with an array of different temperaments. Indeed, part of the enduring appeal of dogs is that many owners feels there is a particular breed for them. This book is a top-to-tail guide to picking your favourite pet, and a celebration of dog breeds and their diversity. It portrays more than 120 of the world's most popular breeds in unparalleled detail, allowing you to admire at leisure or browse with a view to choosing your perfect pet.

You can quickly discover the types of breed that suit you. Firstly, they are arranged in order of size, starting with the smallest, and progressing to the largest, based on official judging standards for each breed. This will serve to give you a key insight into whether a breed potentially could fit easily into your home! Where two breeds are of identical size, the lighter one is listed first. Remember that the height of dogs (as with other livestock) is traditionally measured to the top of the shoulder, rather than to the top of the head, so they are always somewhat taller than indicated.

It is possible to gain a clear insight into the type of dog that is being portrayed, because each entry has its own group indicator as well, revealing whether it is a terrier or a toy breed for example. This is something else to bear in mind, as this is likely to impact directly on the personality and level of activity of a particular breed, influencing whether it may be suitable for your lifestyle.

There is also coverage of the most popular so-called 'designer breeds', which have grown significantly in numbers during the present century, although they cannot be shown in the same way as traditional breeds. In addition, every entry is accompanied by the essential information you require if you are looking for a pet. A silhouette graphic shows instantly how large a puppy will grow. The possible colour combinations for each breed are listed alongside grooming and exercise requirements. Key symbols reveal the needs of each breed, so that quick comparisons can easily be made.

Once you have narrowed down your search, you can study the detailed individual descriptions and learn more about the ancestry of a breed that interests you. It is only over the last 100 years or so that dogs have started to leave their working ancestries behind, with the majority now being kept as household companions. It is therefore important to be aware of the background of a particular breed, because this will still have a significant bearing on its temperament today, and on its behaviour as your pet.

Whether you want to browse through the wonderful world of dog breeds, pick a selection of breeds that you wish to consider as part of your family, or study the details of your chosen dog, *The Right Dog for You* will provide you with a comprehensive view.

PICKING A PET

Picking a purebred or indeed, a designer dog is a decision that needs to be considered very carefully, and it must never be rushed. You will be selecting a companion who, with luck, will be part of your daily life for well over a decade, so you need to be certain that you are making the right choice.

Never be tempted to select a dog purely on the basis of its looks, even if appearance is a good starting point for your deliberations. Delve into the ancestry of a breed that appeals to you, and you will gain the best insight into its personality. This is the ideal way to begin studying many vital considerations, such as how much exercise your new pet will need, and how easily your puppy can be trained. In this way, you can be sure of finding a pet to suit your lifestyle.

There are a number of key factors about your home life that will affect your choice. You must have enough time to care for your dog's needs on a daily basis. If you live in a small flat with no access to a garden and you're out at work all day, then you might need to think about a type of pet that is less demanding than a dog. Dogs must have enough space to exercise and they are highly social by nature, so they will not settle well if they are left alone for long periods, and they may even become destructive around the home when bored.

The size of a breed is, of course, a vital factor. Large dogs have correspondingly bigger appetites and also tend to have shorter lifespans. The grooming needs of the different breeds should also be a consideration. Smooth-coated dogs require relatively little grooming, whereas those with profuse coats will need daily combing and brushing to prevent their fur becoming matted, and some breeds will benefit from regular visits to a dog groomer. So you need to assess how much time you will have to devote to your pet before you make your choice, particularly as some breeds require much more exercise than others.

Right: A puppy can be instantly irresistible, so it is essential that you balance first impressions with careful consideration of the time and care that will be required to give your dog a happy home.

Above: The extremes in size in the dog world are illustrated here by a Great Dane towering over a Papillon. Keeping large dogs can be very demanding. Their dietary requirements are expensive, they need a much more spacious environment than their smaller relatives, and they are usually more demanding in terms of their care.

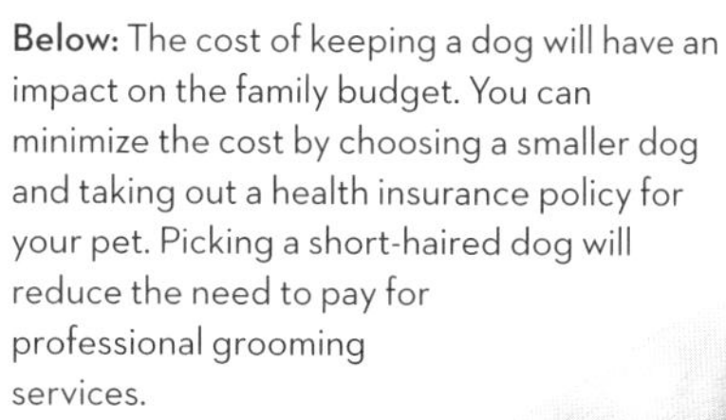

Below: The cost of keeping a dog will have an impact on the family budget. You can minimize the cost by choosing a smaller dog and taking out a health insurance policy for your pet. Picking a short-haired dog will reduce the need to pay for professional grooming services.

Right: Some breeds of dog make much better family pets than others, because they are instinctively friendly by nature. Children who have not had a dog as a pet before need to be taught how to play with their new companion, and never to tease the dog with food or toys.

Above: Once they are properly trained, dogs need to be allowed to run off the leash. Running and playing with your pet is one of the joys of keeping a dog, but do remember that certain breeds will need much more exercise than others.

THE ORIGINS OF TODAY'S BREEDS

Although domestic dogs vary significantly in appearance, they are all descended from the Grey Wolf (*Canis lupus*), which was once among the most widely distributed of all mammals, being found throughout most of the northern hemisphere. The great diversity in form that exists in domestic dogs today was foreshadowed in wolf populations, which showed a marked variance in size and colour depending on their area of origin.

Almost certainly, the domestic dog had several ancestral lines, with domestication occurring in various separate localities at different times. This process began at least 12,000 years ago, and the likelihood is that it may have started much earlier. Sight hounds (dogs that hunt by sight rather than scent) were developed in the vicinity of Egypt and northern Africa, with the Greyhound of today bearing a distinct resemblance to the ancestral form depicted on artefacts dating back thousands of years. Sight hounds probably represent the earliest lineage of the domestic dog, as well as the fastest. They have long legs, relatively narrow heads and broad chests that allow for a good lung capacity.

The large mastiff breeds probably originated in Asia, and were then taken westward along the Old Silk Road into Europe. Meanwhile, the far north was home to the sled breeds, whose descendants (such as the Alaskan Malamute) are still kept for working purposes in the region today. One of the characteristics of these and other so-called spitz breeds from the Arctic area is the shape of the tail, which curls forwards over the back. These dogs have pricked ears and powerful bodies, and a coat that provides them with good insulation against the cold.

Small companion dogs developed at a much later stage, but were certainly quite well-established by the Roman era, some 2,000 years ago. They may well have been the result of selective breeding, with the smallest individuals from litters being mated together, leading to a progressive reduction in the size of their offspring over successive generations.

Up until quite recently, there would not have been such well-defined breeds as there are today. These have only come into existence over the past century or so, as the result of a growing interest in competitive dog showing.

For show purposes, the ideal appearance of each breed is specified by a breed standard. This outlines all the desirable physical features that judges will be looking for, as well as highlighting perceived faults, with dogs in each class being assessed against the standard rather than against each other. Many of the photographs included in this book are of champion dogs, which have been judged to be leading examples of the breeds that they represent.

Left: The ancestry of dog breeds can be traced from the natural magnificence of the Grey Wolf (far left), through the domestication of the dog, to pure-bred examples such as the Basenji (left).

ANATOMY OF THE DOG

Below: Selective breeding has refined the appearance of modern dogs, with each part of the dog being assessed according to the breed standard. The diagram below illustrates the correct terminology for each part of the dog's anatomy.

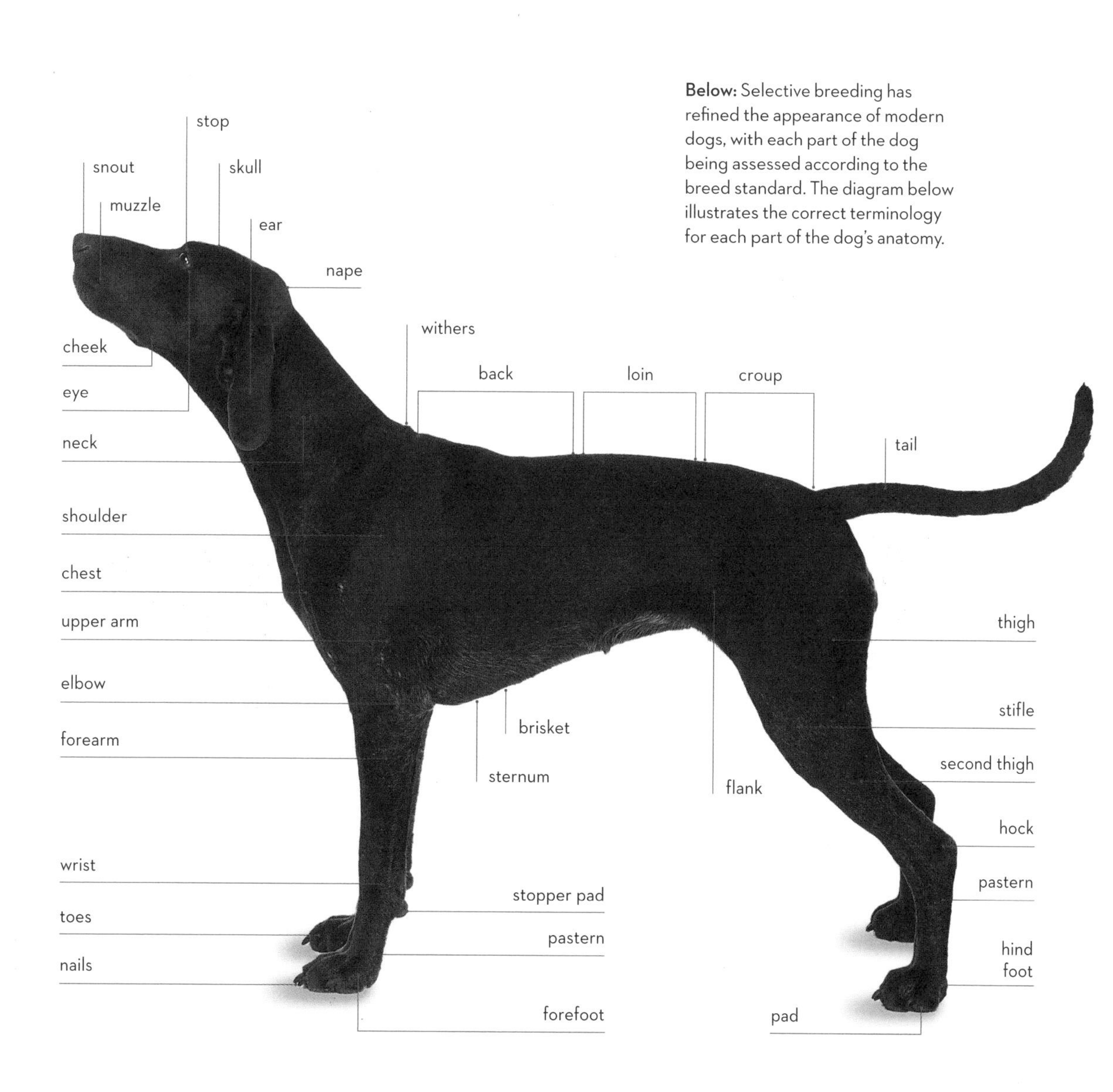

CATEGORIES OF BREEDS

The way in which breeds are categorized is not standardized throughout the world. The Kennel Club in Britain and the American Kennel Club (AKC) have seven divisions, as shown below, while the European FCI (Fédération Cynologique Internationale) has ten divisions.

Fundamentally, all of these are based on the original purpose for which the breed was developed. Anomalies in the listings are most likely to arise in the case of breeds that were created for varying tasks, as with dogs that were developed as farm breeds. These animals may have been used for herding purposes, as guard dogs, and also as hunting companions.

There are now approximately 400 different breeds of dog in the world today. The word 'breed' is used to describe a population of dogs that, when paired together, will produce puppies of corresponding appearance. In a few instances, there may only be a very marginal difference between breeds, as shown by the case of the Norwich and Norfolk Terriers, where the division is based essentially on a variance in their ear carriage. Norwich Terriers have pricked ears, while the ears of Norfolk Terriers hang down the sides of the head.

About half of the breeds in the world today are well-known in show circles. Many of the others are localized. They will be of recognizable appearance or 'type', but there might not always be a specific show standard drawn up for that particular breed.

HERDING - He These dogs have been developed primarily to work with farmstock, and so form a close bond with their owners. Herding breeds are characterized by their natural intelligence and their ability to learn quickly. They are energetic by nature, and will rapidly become bored, frequently proving to be destructive around the home unless they can be given plenty of exercise.

HOUND - H Members of this group represent the oldest category of dogs, having been bred as hunting companions. Based on existing archeological evidence, it seems that their appearance has changed little over the course of thousands of years. The Hound group is usually subdivided into sight hounds, which are the natural sprinters of the dog world, and the slower scent hounds, which often hunt in packs.

NONSPORTING - NS This AKC division effectively represents a miscellaneous group, corresponding to the Utility division adopted by the Kennel Club in Britain. It therefore comprises a diverse group of dogs, which were originally created for a number of different tasks and do not fit conveniently into any of the other groupings. Unsurprisingly, this is therefore the most varied division.

Opposite, below and right: The book is divided into the seven divisions of the American Kennel Club in the order shown here: Herding, Hound, Nonsporting, Sporting, Terrier, Toy and Working. A full description of the main breeds is given in each section, followed by an overview of six further breeds within the division.

While some breeds are rare, and even on the verge of extinction in certain cases, there are still new breeds being created today. The tendency over recent years has been to cross existing breeds and so develop dogs of attractive appearance that will make good companions – although this is often frowned on by purists. The best-known variant of this type developed to date is the Labradoodle – a cross between the Labrador Retriever and the Poodlc. Although such dogs have not yet obtained international recognition in show circles, they are highly sought-after as pets.

SPORTING - S Although these breeds are often described today as gun dogs, their origins in many cases predate the development of shooting as a leisure sport. They are used to locate, flush and then retrieve game after it has been shot. These dogs are easily trained and work well alongside people, and so have become popular as pets. However, it is essential that they have plenty of exercise.

TERRIER - Te Members of this group are relatively small, rather independently minded dogs. The group was largely developed in Britain, where such dogs were often sent underground to flush out foxes and other creatures, which could then be pursued by hounds. Terriers were also popular on farms for their rodent-killing abilities, while their boldness led to a period when certain breeds were used for dogfighting.

TOY - T A group characterized by their small size, toy dogs were created primarily as companions. Many of today's most popular breeds were originally bred as the pets of wealthy ladies around the royal courts of Europe. Some, like the Italian Greyhound, are effectively scaled-down versions of larger breeds, whereas others are quite distinct in appearance. They are affectionate, friendly dogs.

WORKING - W Many members of this group were originally bred as guardians of flocks and property, and so are not necessarily friendly by nature. They are generally large and powerful dogs, and must be firmly trained. There are some, however, whose strength has been used for more gentle purposes, such as pulling carts or helping fishermen.

DESIGNER DOGS

Dogs of this type have become increasingly popular over recent years. They are basically cross-breeds, created from two different breeds, with many but not all having a part-poodle ancestry. The appearance of puppies within a litter can vary significantly as a result. Their names provide an insight into their origins, typically representing a combination of that of both parent breeds, as illustrated by the Puggle (a Pug x Beagle cross-bred) for example.

The origins of this group date back to an Australian attempt to create a hypoallergenic guide dog in 1989 by combining a Labrador Retriever with a Standard Poodle. The theory was that the poodle input would help to prevent shedding of hair in the resulting puppies, which became known as Labradoodles. They now rank as one of the most popular designer dogs worldwide today.

There are a number of misconceptions surrounding designer dogs, however, not least the belief that they are healthier than purebred dogs. Furthermore, in the cases of breeds with a poodle ancestry, there is also no guarantee that all the puppies in a litter will not shed, or indeed that they will be hypoallergenic. It is now clear that some people are sensitive to the tiny fragments of shed skin, known as dander, rather than the hair itself.

Today, a significant part of the appeal of a designer dog for many owners is the fact that they will be acquiring a pet with some recognizable attributes, and yet an individual appearance at the same time. Even so, in any litter there are likely to be individuals that err more towards one or other parent breed in appearance, rather than being consistently midway between them in this regard. Similarly, in terms of temperament, some variability can be anticipated.

Obtaining a designer dog is rarely a cheap option. Today, puppies of this type tend to be more expensive than those of either of the parent breeds. It may also be more difficult to acquire a puppy of the less-common crosses, and you will need to be alert, regularly checking online sites listing dogs for sale, as well as probably being prepared to travel some distance when seeking the pet of your choice. The recognition 'D' is used for designer dog entries later in the book.

Far left: In the case of Labradoodles, pale apricot and also black colouration (as in the dog behind) are both commonly seen, given that similar colours are associated with the parent breeds.

Left: Goldendoodles are known for being energic and playful. They love going on long walks, runs and hikes with their owners – the perfect companions for active families.

Left: A Cockapoo with a relatively long coat. As a general rule, the coat of an individual designer dog will inevitably be at its shortest when it is a young puppy. However, littermates may vary significantly in this respect.

Right: The appearance of a designer dog may often more closely resemble one of its parent breeds, as shown here. The young chug (a Chihuahua x Pug) looks not that dissimilar from the older male Pug on the right, although it lacks the distinctive curled tail associated with that breed.

USING THE PET SPECS

All the various components on the Pet Spec pages have been carefully designed both to facilitate easy access to the relevant information and to assist in easily comparing the needs of different breeds.

The photographs taken from two different angles give a much clearer indication of the overall appearance of the dog than the standard profile seen in most dog books, as features such as the width and depth of the chest and the breadth of the body are clearly apparent.

In all cases, the symbols in the 'Overview' section refer to a young, fit adult dog. It is important to bear in mind that as dogs grow older, so their dietary and exercise needs are likely to decrease.

The 'Child friendliness' rating is based on children who live with the dog. A house pet is often likely to be less inclined to play with other children visiting the home, and may bark at them initially, viewing them as strangers. This can cause problems, and you should always be cautious when allowing your pet to mix with your children's friends, particularly if they are nervous of dogs.

A number of hereditary and congenital health problems may occasionally afflict different breeds. Some, such as hip dysplasia, are much more common among some breeds than others.

The Specification will provide you with a very good starting point in choosing your new dog. However, if you have any concerns at all about your pet, you should always seek the expert advice of a vet at an early stage.

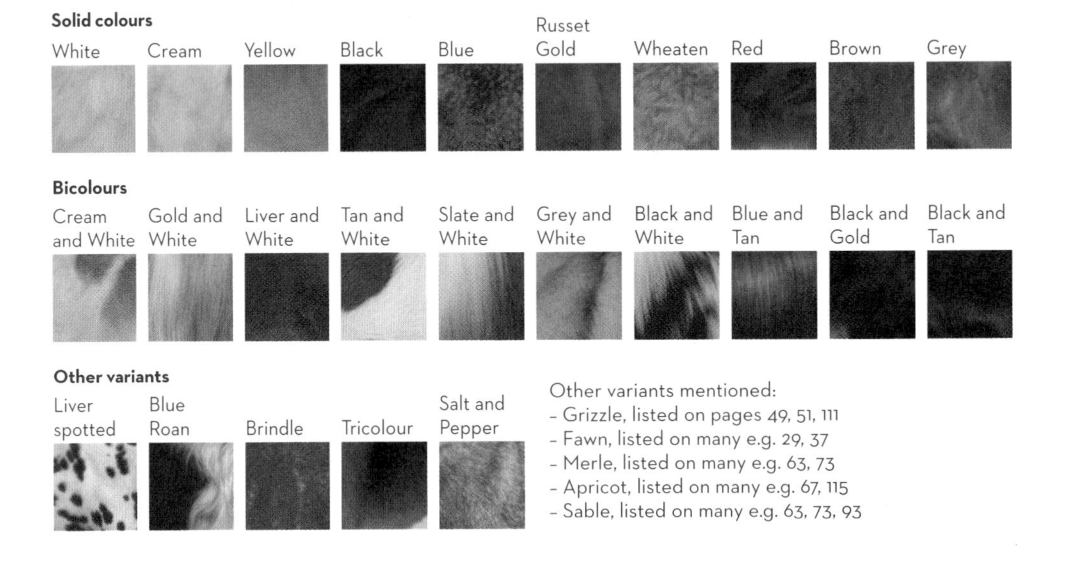

How to use this book

1 Breed The breeds in this book are arranged in order of size, starting with the smallest and ended in the largest, then followed by the designer dogs.

2 Recognition The AKC system is used in this book and the group abbreviations appear too. This confirms how widely the breed in question is recognized for show purposes. In some cases, where there is more than one official registry in a particular part of the world, this is no guarantee that all bodies will recognize the breed or classify it in the same group.

3 Features The vital statistics of each breed show the proportions of head, eyes, ears, chest, tail and bite, which collectively contribute to its "type" or appearance. Also listed are height (in the image caption), measured to the highest part of the shoulder, and ideal weight.

4 Symbols A rating of 1–3 is given, with 1 meaning least and 3 the most. A breed with 2 children, 1 brush, 1 bowl and 2 running dogs is reasonably child-friendly, needs minimal grooming, has a small appetite and has average exercise requirements.

5 Colour The acceptable range of colours associated with a particular breed is highlighted here, along with mention of specific markings that may be required. A colour reference, although not exhaustive, for the solid, bicolour and most other variant shades is shown below left.

6 Images The photos show the same dog, in a show stance, from two different angles, allowing you to take a virtual tour of each breed's features.

7 Group The breed entries are divided into seven distinct groups, relating to their original function and their show categorization. These are Herding (He), Hound (H), Nonsporting (NS), Sporting (S), Terrier (Te), Toy (T), Working (W) and Designer (D).

8 Text entry The entry for each breed features an introduction, and then concise descriptions of origin, appearance, personality, health and care, owner requirements, and finally a feature suggesting similar breeds that might appeal.

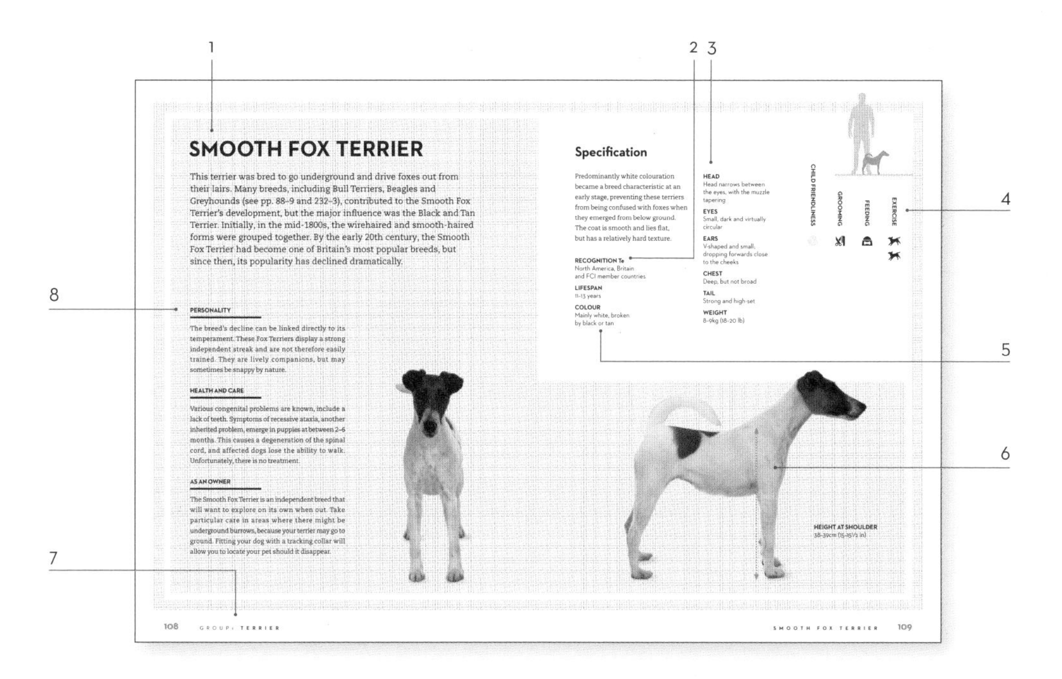

SMOOTH FOX TERRIER

This terrier was bred to go underground and drive foxes out from their lairs. Many breeds, including Bull Terriers, Beagles and Greyhounds (see pp. 88–9 and 232–3), contributed to the Smooth Fox Terrier's development, but the major influence was the Black and Tan Terrier. Initially, in the mid-1800s, the wirehaired and smooth-haired forms were grouped together. By the early 20th century, the Smooth Fox Terrier had become one of Britain's most popular breeds, but since then, its popularity has declined dramatically.

PERSONALITY

The breed's decline can be linked directly to its temperament. These Fox Terriers display a strong independent streak and are not therefore easily trained. They are lively companions, but may sometimes be snappy by nature.

HEALTH AND CARE

Various congenital problems are known, include a lack of teeth. Symptoms of recessive ataxia, another inherited problem, emerge in puppies at between 2–6 months. This causes a degeneration of the spinal cord, and affected dogs lose the ability to walk. Unfortunately, there is no treatment.

AS AN OWNER

The Smooth Fox Terrier is an independent breed that will want to explore on its own when out. Take particular care in areas where there might be underground burrows, because your terrier may go to ground. Fitting your dog with a tracking collar will allow you to locate your pet should it disappear.

Specification

Predominantly white colouration became a breed characteristic at an early stage, preventing these terriers from being confused with foxes when they emerged from below ground. The coat is smooth and lies flat, but has a relatively hard texture.

RECOGNITION Te
North America, Britain and FCI member countries

LIFESPAN
11–13 years

COLOUR
Mainly white, broken by black or tan

HEAD
Head narrows between the eyes, with the muzzle tapering

EYES
Small, dark and virtually circular

EARS
V-shaped and small; dropping forwards close to the cheeks

CHEST
Deep, but not broad

TAIL
Strong and high-set

WEIGHT
8–9kg (18–20 lb)

CHILD FRIENDLINESS
GROOMING
FEEDING
EXERCISE

HEIGHT AT SHOULDER
38–39cm (15–15½ in)

108 GROUP: TERRIER

SMOOTH FOX TERRIER 109

IN-DEPTH PROFILES OF BREEDS, ARRANGED BY ADULT SIZE

CHIHUAHUA

Named after the Mexican province where it originated, the Chihuahua's origins are mysterious. It is probably descended from dogs kept by the ancient Aztecs, and ranks as the world's smallest dog today. Brought to the USA in the mid-1850s, the early examples had longer faces and larger, more batlike ears than contemporary Chihuahuas.

PERSONALITY

Lively and with a surprisingly loud bark, the Chihuahua possesses a bold, fearless nature. These dogs form close bonds both with each other and with their owners. They may sometimes shiver with excitement, although the shorthaired variety is also sensitive to the cold.

HEALTH AND CARE

Chihuahuas can suffer from a range of inherited ailments, some of which – such as hydrocephalus, which causes an abnormally swollen head – will be evident at birth. They must not become fat, because they are then vulnerable to a narrowing of the airway to the lungs.

AS AN OWNER

Grooming of the smooth-coated form is very straightforward. Check the mouths of puppies to ensure they lose their milk teeth, because these are sometimes retained as the permanent teeth emerge. Chihuahuas can be fussy eaters.

Specification

Its large, domed head has a vulnerable, unprotected gap at its centre, known as a molera, meaning that these tiny dogs must be handled carefully. The traditional form is smooth-coated, with sleek, glossy fur, while the longhaired variety has an equally soft, sometimes slightly curly coat, a long plume of hair on the tail and fringed ears.

RECOGNITION T
North America, Britain
and FCI member countries

LIFESPAN
12–15 years

COLOUR
Any colour or pattern

HEAD
Characteristic apple-domed skull and a short nose

EYES
Well-spaced and round

EARS
Large and positioned at an angle of about 45 degrees on the sides of the head

CHEST
Prominent with a deep brisket

TAIL
Medium-length and carried over the back; curled and tapered to a point

WEIGHT
1–1.75kg (2–4 lb) preferred, but can be up to 2.75kg (6 lb)

CHILD FRIENDLINESS

GROOMING

FEEDING

EXERCISE

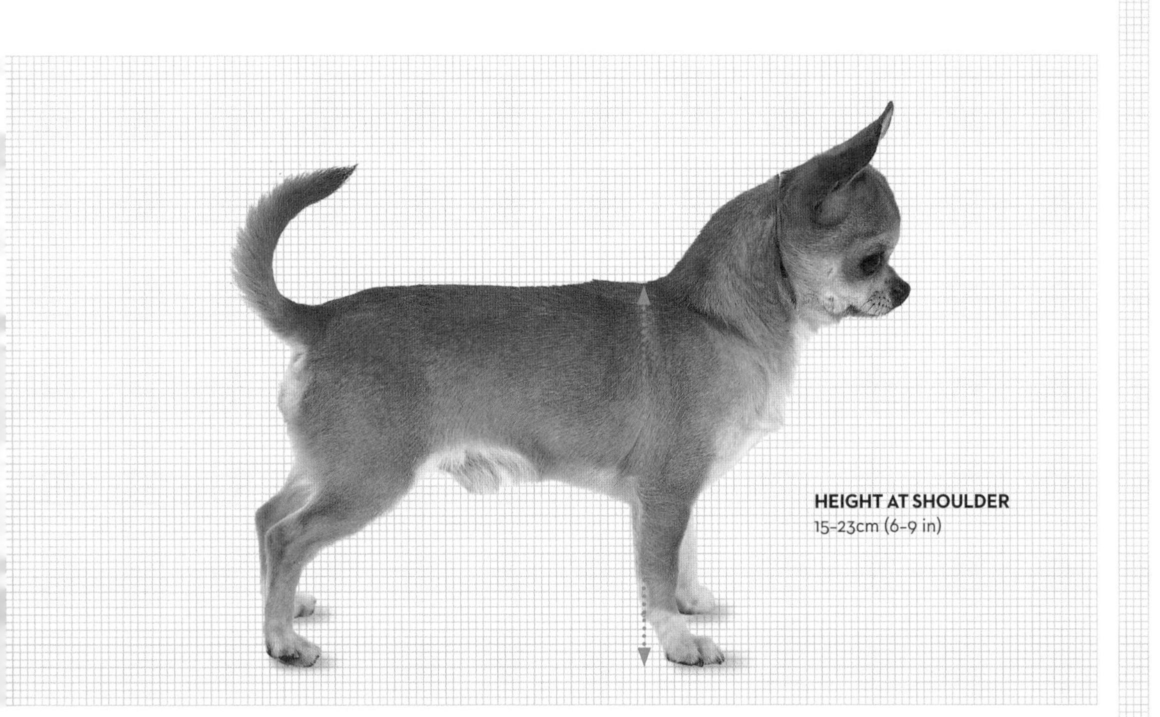

HEIGHT AT SHOULDER
15–23cm (6–9 in)

GRIFFON BRUXELLOIS

Also sometimes known as the Brussels Griffon, these small terrier-like dogs have been popular in their Belgian homeland for more than 600 years. Griffons Bruxellois were originally kept by the aristocracy, but soon gained wider popularity. Living around stables, they proved to be proficient ratters. They used to be more terrier-like in appearance, but during the 1800s, crossbreeding with the Pug flattened their facial shape (in addition to creating the smooth-coated Brabançon form).

PERSONALITY

Lively and alert, these dogs are playful and easy to train, but have a rather sensitive nature. However, leash-training can sometimes prove difficult.

HEALTH AND CARE

The fertility of the breed has fallen significantly over the past century. In the case of bitches that become pregnant, the birth can be difficult, too, and litters quite frequently consist of just a single puppy, which may need to be born by Caesarean section. The wire coat requires hand-stripping to keep it looking neat and tidy.

AS AN OWNER

The Griffon Bruxellois thrives on attention and will form a close bond with members of its immediate family. A Griffon is a good choice if you have a cat, since these small dogs usually live in harmony with a feline companion. This breed is less suitable for a home where there are young children because they do not enjoy boisterous play.

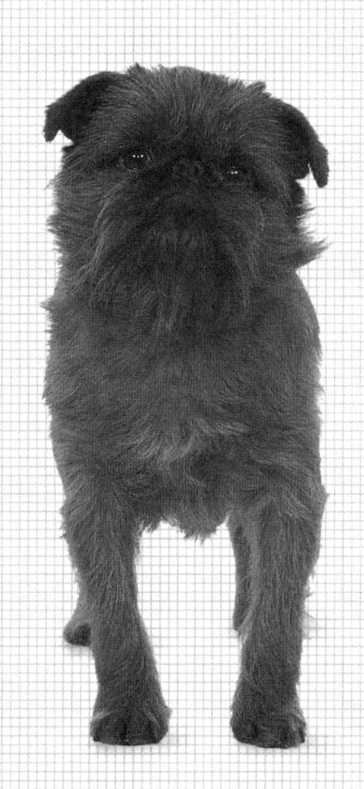

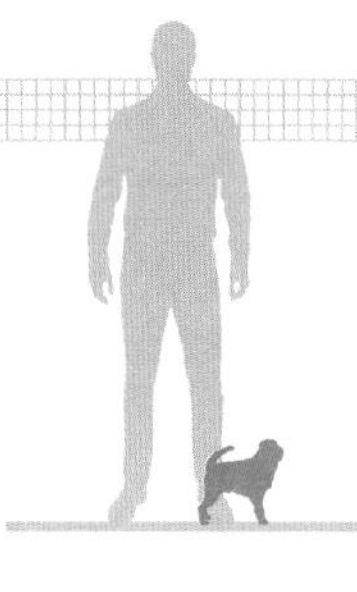

Specification

The coat has a harsh, wiry texture, being longer on the head. The only white allowed is so-called 'frosting' on the muzzle.

RECOGNITION T
North America, Britain and FCI member countries

LIFESPAN
12–15 years

COLOUR
Predominantly red; black; black and tan; *belge* (black and reddish-brown)

HEAD
Large and round, and wide between the ears

EYES
Large, round and dark in colour

EARS
Small and set high; held semi-erect

CHEST
Relatively deep and wide

TAIL
Angled at 90 degrees from the back, usually carried high

WEIGHT
Ideally 3.5–4.5kg (8–10 lb); must not exceed 5.5kg (12 lb)

HEIGHT AT SHOULDER
20cm (8 in)

DACHSHUND

French noblemen fleeing from the French Revolution of 1793 brought bassets of various types with them to the area of present-day Germany. These were crossed with native German dogs, and so Dachshunds evolved. They may not look like traditional hounds, but these short-legged German dogs were evolved to hunt badgers underground, while the miniature forms hunted rabbits.

PERSONALITY

Bold and fearless in spite of their small stature, Dachshunds are loyal and good-natured towards family members. They are alert watchdogs, with a more intimidating bark than their size might suggest.

HEALTH AND CARE

Dachshunds are vulnerable to intervertebral disc problems, linked to their profile. They must not be encouraged to climb stairs or jump onto chairs for this reason. They must also not be allowed to become overweight, as this will place extra strain on the vertebral column. Shorthaired Dachshunds are susceptible to skin mites, which will cause extensive hair loss.

AS AN OWNER

Choose the coat type carefully, bearing in mind that a long-haired Dachshund requires more grooming than either of the other variants. The breed does not need spacious surroundings and is suitable for urban areas where there are parks nearby for exercise. Dachshunds must be exercised on a harness rather than a collar to protect their vulnerable backs.

Specification

The standard, smooth-haired Dachshund is the original form. The longhaired variety arose as the result of crossbreeding with spaniels, and the Wirehaired German Pinscher probably contributed this characteristic to the breed. Dachshunds were then scaled down, creating corresponding miniature varieties.

RECOGNITION H
North America, Britain and FCI member countries

LIFESPAN
12–13 years

COLOUR
White only on the chest

HEAD
Tapers to the tip of the nose

EYES
Almond-shaped and of medium size

EARS
Rounded and set close to the top of the head

CHEST
Prominent breastbone with a depression on each side

TAIL
Carries on without greater curvature at the end of the body

WEIGHT
7–14.5kg (16–32 lb)

CHILD FRIENDLINESS

GROOMING

FEEDING

EXERCISE

HEIGHT AT SHOULDER
Not specified; typically 20–26.5cm (8–10½ in)

SHIH TZU

The name of this ancient Oriental breed means 'Lion Dog', and it is traditionally pronounced 'sher-zer'. The way in which the hair grows in flowing waves on its head explains why it has also been called the Chrysanthemum Dog. The Shih Tzu is the result of crossbreeding between Lhasa Apso and Pekingese (see pp. 30–1) at the Imperial Palace in Beijing. The breed remained unknown in the West until the 1930s.

PERSONALITY

The Shih Tzu has inherited a playful side to its nature from its Pekingese forebears, but shares their dignified temperament, too. These dogs make friendly, affectionate companions.

HEALTH AND CARE

A hereditary malformation of the kidneys can sometimes afflict the Shih Tzu, but symptoms will not become apparent until young dogs are at least a year old. Another hereditary problem may be various abnormalities affecting the blood clotting system. The first visible symptom of this can be the presence of blood blisters, which may become large, under the skin.

AS AN OWNER

The Shih Tzu demands considerable grooming time because of its long, dense, double-layered coat. The hair extending up from the nose is usually tied up on the head in a topknot. This breed is suitable for flat dwellers, because it is content living in a fairly limited space.

Specification

The head shape of the Shih Tzu reflects its ancestry. It has a more rounded skull than the Lhasa Apso, but a longer, more prominent muzzle than the Pekingese, and its eyes are clearly visible. Unusually however, the hair on its nose grows up towards the forehead.

RECOGNITION T
North America, Britain and FCI member countries

LIFESPAN
11–13 years

COLOUR
No restrictions

HEAD
Broad and round

EYES
Well-spaced and dark in colour

EARS
Large and hang down the sides of the head

CHEST
Deep and broad, with rib cage reaching just below the elbow

TAIL
Set high and curved forwards over the back

WEIGHT
4–7kg (9–16 lb)

CHILD FRIENDLINESS

GROOMING

FEEDING

EXERCISE

HEIGHT AT SHOULDER
20–28cm (8–11 in); ideally 23–26.5cm (9–10½ in)

DANDIE DINMONT TERRIER

This is the only breed of dog that owes its name to a character in a popular novel. Sir Walter Scott featured Dandie Dinmont, the owner of one of these terriers, in *Guy Mannering*, published in 1814. The breed became known as Dandie Dinmont's Terrier, and this was ultimately shortened to its current name. It was bred from native terriers in the southern borders of Scotland during the early 1700s, but its origins are unclear.

PERSONALITY

This breed is intelligent but reserved, although it does have a more extrovert, playful side. Tenacious and bold, it is not easily intimidated even by dogs that are much larger dogs.

HEALTH AND CARE

The long body makes this breed susceptible to intervertebral disc problems. These discs normally act as cushions between the vertebrae, but they can become displaced. Often, however, careful nursing and treatment will lead to recovery. Professional grooming is considered important to maintain the Dandie Dinmont's unique appearance.

AS AN OWNER

Aim to minimize the risk of any back problems arising. Exercise your pet using a harness instead of a collar to reduce pressure on the vulnerable neck region, and discourage it from jumping up on furniture or climbing stairs. Use a stair guard if necessary.

Specification

The Dandie Dinmont has a long, low-slung body and a topknot of silky hair on the head. The neck is muscular, suitable for a hunting breed that had to cope with overpowering badgers and other big animals. Its colouring is either 'salt and pepper' – shades between silvery-grey and blue-black – or 'mustard', extending from fawn to reddish brown.

RECOGNITION Te
North America, Britain and FCI member countries

LIFESPAN
11–13 years

COLOUR
Mustard or pepper

HEAD
Broad skull, with a well-domed forehead and a deep, powerful muzzle

EYES
Large, round and widely spaced

EARS
Long, widely spaced and set well back

CHEST
Well-developed, reaching between the forelegs

TAIL
Carried in a scimitar-like curve

WEIGHT
8–11kg (18–24 lb)

GROOMING

FEEDING

EXERCISE

HEIGHT AT SHOULDER
20–28cm (8–11 in)

PEKINGESE

Named after the Chinese city of Peking (now Beijing), Pekingese were highly prized, with ownership being restricted to the Emperor of China. These dogs were cared for by eunuchs, and stealing or selling one was punishable by death. They used to be known as Sleeve Dogs: being smaller in those days, they could be carried in the sleeves of the flowing robes of Chinese courtiers. Queen Victoria, a great dog lover, was presented with the first Pekingese seen outside China in the 1860s.

PERSONALITY

Loyal yet stubborn and courageous, the Pekingese has a confident and independent nature.

HEALTH AND CARE

Daily grooming is essential. The Peke's prominent eyes are susceptible to injury, and the compacted shape of the breed's face means that these dogs are prone to tearstaining of their fur. The long body of the Pekingese also means that it is at risk from back injuries, and they must not be encouraged to jump or to climb stairs.

AS AN OWNER

Do not be fooled by the appearance of the Pekingese. It possesses good stamina, although never exercise these dogs in hot weather, especially around the middle of the day, when they will be most vulnerable to the effects of heatstroke.

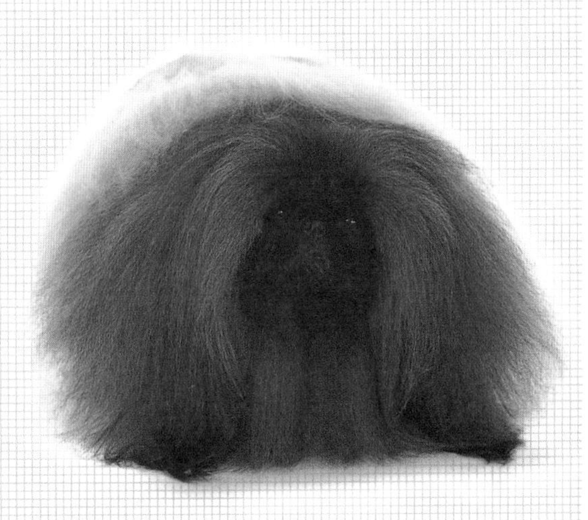

Specification

Pekingese have compact faces with very prominent eyes. The coat is now more profuse than in the past, trailing to the ground, and their legs are shorter than those of their ancestors. They retain a long area of hair, resembling a mane, around the neck, which is why they used to be called Lion Dogs.

RECOGNITION T
North America, Britain and FCI member countries

LIFESPAN
10–12 years

COLOUR
Any colour or pattern; commonly golden-red, with such individuals described as Sun Dogs

HEAD
Massive and broad, with a short nose.

EYES
Large, prominent and dark

EARS
Heart-shaped, and lie close to the head

CHEST
Broad, with well-sprung ribs

TAIL
Set high on the back and slightly curved over it

WEIGHT
Up to 6.5kg (14 lb)

GROOMING

FEEDING

EXERCISE

HEIGHT AT SHOULDER
Up to 23cm (9 in)

JAPANESE CHIN

In terms of its behaviour, the Japanese Chin has been likened to a cat in some respects, being able to climb well and proving a nimble jumper. It is a very ancient breed, whose origins date back over a millennium, and probably shares a common ancestry with the Pekingese (see pp. 30–1). It was a favoured pet of the Japanese nobility, first seen in Europe in the 1600s, and known initially as the Japanese Spaniel in North America.

PERSONALITY

While forming a close bond with people it knows well, the Japanese Chin is shy with strangers. It is not an effective guard dog though, because it rarely barks.

HEALTH AND CARE

Although usually healthy, the Japanese Chin may be inclined to snore as a result of its face shape. Its puppy coat is noticeably shorter than that of an adult, but accustom a young dog to daily grooming from an early stage.

AS AN OWNER

This breed is ideal if you live in a flat, because it is one of the few dogs that is content in these surroundings, provided that it can be taken out as necessary during the day. Training is straightforward, as the Japanese Chin's intelligence means that it learns quickly, although members of the breed can be stubborn too, but they respond well to positive encouragement.

Specification

A flattened face and an upturned nose are characteristic, with the ears hanging far down the sides of the head. The coat is long and silky. Feathering occurs on the back of the front legs, with the feet being long and narrow. A clear white blaze between the eyes is desirable.

RECOGNITION T
North America, Britain and FCI member countries

LIFESPAN
11–13 years

COLOUR
Bicolour – black and white or red and white

HEAD
Large and rounded, not domed

EYES
Well-spaced, large and dark

EARS
Widely spaced and set high

CHEST
Wide, contributing to its stocky appearance

TAIL
Set high; turned from the base or carried down over the side

WEIGHT
1.75–3kg (4–7 lb)

CHILD FRIENDLINESS

GROOMING

FEEDING

EXERCISE

HEIGHT AT SHOULDER
23cm (9 in)

YORKSHIRE TERRIER

The Yorkie, as it is affectionately known, has become one of the most popular terriers worldwide, being kept as a companion breed rather than for working purposes. It is well-suited to urban living, with its origins tracing back to the mill towns of Yorkshire in northern England, where its ancestors were used to hunt rats and mice. Various terrier breeds in the region probably contributed to its ancestry, including the Manchester Terrier and the now-extinct Leeds Terrier.

PERSONALITY

This is a dog with a large personality in spite of its diminutive size, the Yorkshire Terrier has a bossy yet bold nature. It is also fearless and, in spite of its pampered appearance, it has not forgotten its rodent-killing past.

HEALTH AND CARE

Unfortunately, an inherited weakness of the kneecaps, known as patellar luxation, is relatively common and will require surgical correction. This is indicated by signs of lameness in puppies when they are 4–6 months old. Young Yorkshire Terriers normally have black areas on the coat which will turn to the desired characteristic steel-blue colouration over time.

AS AN OWNER

This is a breed that needs plenty of grooming. The area around the mouth must be kept clear of food, being cleaned and combed regularly.

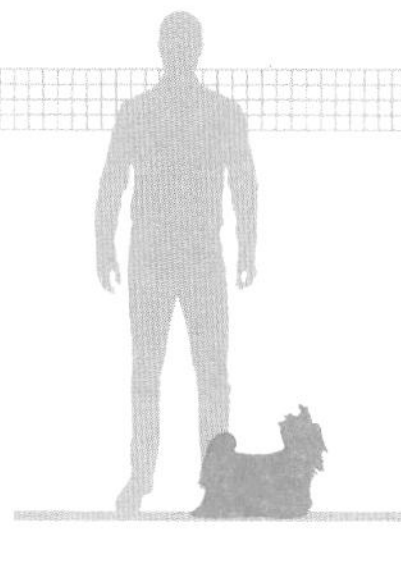

Specification

The coat of the Yorkshire Terrier has a fine texture with an attractive gloss and should extend down each side of the body to ground level. The hair on the head is often tied in the centre with a bow, or parted to the sides and held in place with two bows. Hair on the tips of the ears is trimmed short.

RECOGNITION T
North America, Britain and FCI member countries

LIFESPAN
11–13 years

COLOUR
Blue and tan

HEAD
Relatively small and with a muzzle of matching length

EYES
Medium-sized and dark in colour

EARS
Small, V-shaped and erect or semi-erect

CHEST
Moderate, separating straight front legs

TAIL
Carried above the level of the back

WEIGHT
3kg (7 lb)

CHILD FRIENDLINESS

GROOMING

FEEDING

EXERCISE

HEIGHT AT SHOULDER
23cm (9 in)

SILKY TERRIER

This small Australian dog was created as a companion breed during the late 1800s, being known originally as the Sydney Silky Terrier, but it is now called the Australian Silky Terrier. Its development was centred on the southeastern states of Victoria and New South Wales, particularly in the vicinity of Sydney. It was bred from a combination of Yorkshire Terrier and Australian Terrier crosses, which were originally carried out to improve the colouring of the latter's blue and tan coat. Skye Terriers were probably responsible for the breed's pricked ears.

PERSONALITY

Silky Terriers possess the character typical of a larger terrier; they are lively and alert, and make good guard dogs. They are also friendly and trusting with their owners.

HEALTH AND CARE

The Silky Terrier is a robust breed, in spite of its diminuitive or reduced size. Its dark nails may need regular clipping. One of the advantages of the Silky Terrier in the home is that it does not moult.

AS AN OWNER

Although popular in Australia, the Silky Terrier is not frequently seen elsewhere. It is particularly uncommon in Britain, where Yorkshire Terriers have retained their popularity. As its name suggests, the coat has a wonderful silky texture, which is most profuse in adult dogs.

Specification

The coat is particularly distinctive, having a silky texture and parting down the midline of the body. These terriers have a fine-boned appearance, but they are not delicate. Silky Terriers are always blue and tan, but the depth of the blue colouring varies from a silvery shade to a dark slate blue.

RECOGNITION T
North America, Britain and FCI member countries

LIFESPAN
11–13 years

COLOUR
Blue and tan, with a lighter silvery or fawn topknot on the head

HEAD
Medium-length and wide between the ears

EYES
Round, small and dark

EARS
Small, V-shaped pricked ears; set high

CHEST
Medium both in depth and breadth

TAIL
Held erect and clearly visible

WEIGHT
3.5–4.5kg (8–10 lb)

CHILD FRIENDLINESS

GROOMING

FEEDING

EXERCISE

HEIGHT AT SHOULDER
23cm (9 in)

BICHON FRISE

The Bichon Frise's name is pronounced 'bee-shon Free-zay.' 'Bichon' is a contracted form of Barbichon, meaning 'small Barbet' (a French breed of water spaniel), while 'Frise' refers to the dog's soft curly coat. The Bichon Frise became popular as a ladies' companion in royal circles, first in France during the 1500s and then in Spain. It fell out of favour in the 1800s, and not until the mid-1950s did it became popular again.

PERSONALITY

A small dog with an appealing nature, the Bichon Frise is a natural entertainer and clown that thrives on attention. This is why it became a popular choice with street performers when it fell out of favour in royal circles.

HEALTH AND CARE

Like many other toy breeds, the Bichon Frise sometimes suffers from a weakness of the kneecaps, which is likely to require surgical correction. Epilepsy is also a problem in some bloodlines. The Bichon Frise's coat needs to be trimmed regularly to maintain its distinctive powder-puff appearance.

AS AN OWNER

The Bichon Frise has a playful, gentle nature, proving very suitable for a home with younger children. Grooming care such as trimming is likely to require professional help.

Specification

The attractive dark eyes of the Bichon Frise are emphasized by dark matching rims, contrasting with the snow-white coat. The soft thick under layer is overlaid by a curly, coarser top coat, creating a springy texture that feels rather like velvet. The Bichon Frise is trimmed to show off its physique, with longer furnishings on the head and on the tail. The pads are black, as is the nose.

RECOGNITION T
North America, Britain and FCI member countries

LIFESPAN
11–13 years

COLOUR
White

HEAD
Skull longer than the muzzle

EYES
Medium-sized, relatively round and dark

EARS
Narrow and hang closely down the sides of the head

CHEST
Deep brisket and well-developed

TAIL
Carried raised; curled loosely over the back

WEIGHT
3–5.5kg (7–12 lb)

CHILD FRIENDLINESS

GROOMING

FEEDING

EXERCISE

HEIGHT AT SHOULDER
23–30.5cm (9–12 in)

LHASA APSO

This breed originated in one of the remotest areas in the world, living in isolation for thousands of years. In its Tibetan homeland, the Lhasa Apso's name translates roughly as 'hairy lion dog', referring both to its profuse coat and its distinctive character. Lhasa Apsos were probably developed by monks from small Tibetan Terrier stock, and were regarded as repositories for the souls of dead monks. They were therefore viewed as sacred, given only rarely to outsiders. Some reached England during the early 20th century, and the first seen in North America were presented as a gift by the then Dalai Lama in 1933. It has now become a very popular companion breed.

PERSONALITY

The Lhasa Apso is quiet and can prove playful, as well as being loyal. However, the breed also has a stubborn streak.

HEALTH AND CARE

Inguinal hernias are common, usually requiring surgical correction. Occasionally, puppies suffer from a brain disorder known as lissencephaly, causing a loss of coordination. This disorder is most common in certain bloodlines.

AS AN OWNER

Grooming will be time-consuming. The breed's long coat is traditionally parted down the middle of the back and combed smoothly down to the ground.

Specification

The Lhasa Apso's appearance has been shaped over the centuries by the landscape in which it evolved. Its long body provides good lung capacity, which is vital in the thin mountain air, and the dense coat offers good protection against the bitter cold of the night in this region. The thick covering of fur on its pads cushions against the rough, stony ground.

RECOGNITION NS
North America, Britain and FCI member countries

LIFESPAN
11–13 years

COLOUR
Any colour

HEAD
Relatively narrow skull, with a medium-length muzzle

EYES
Medium-sized

EARS
Hang down over the sides of the face, with long feathering

CHEST
Medium, but with good rib cage

WEIGHT
6–7kg (13–15 lb)

GROOMING

FEEDING

EXERCISE

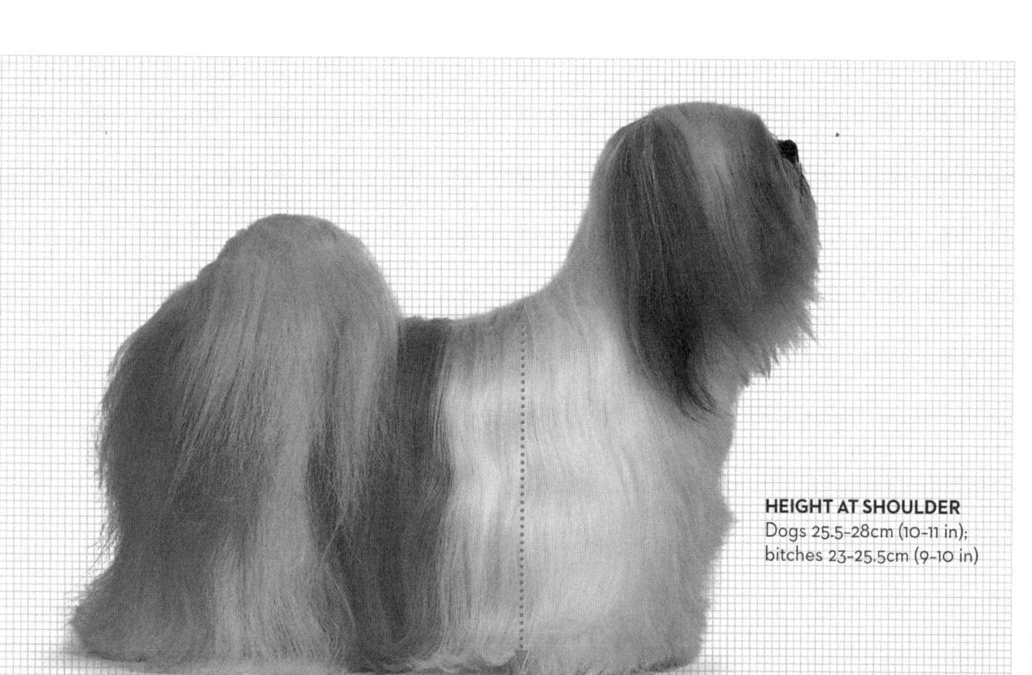

HEIGHT AT SHOULDER
Dogs 25.5–28cm (10–11 in); bitches 23–25.5cm (9–10 in)

CAIRN TERRIER

The Cairn is named after the cairns, or piles of stones that were built up along Scottish roadsides, both to mark boundaries and for ceremonial purposes. These provided ideal retreats for rats and even foxes, and farmers used terriers to drive these unwelcome visitors out. The breed's forerunners were already being kept in Scotland's Western Highlands and on the Isle of Skye over 400 years ago, and the Cairn's appearance has changed remarkably little over this period. Until 1909, it was called the Shorthaired Skye Terrier.

PERSONALITY

A typical terrier, with a bold, inquisitive nature and a strong character, the Cairn delights in human company, but is less keen on canine companionship. In spite of its size, this terrier is hardy and will expect to go out even in bad weather.

HEALTH AND CARE

The breed suffers from a number of genetic problems, including haemophilia. Puppies aged 4–7 months can suffer from craniomandibular osteopathy, affecting the jaws and adjacent tympanic bullae. This raises the puppy's temperature and makes eating painful. With treatment, they will make a full recovery.

AS AN OWNER

The Cairn is a sturdy breed, with considerable energy, so although it may not run hard unless it is chasing or hunting, it will trot alongside you for considerable distances. If it does find something to chase, there is a risk that it may disappear underground in pursuit of its quarry or plunge into water.

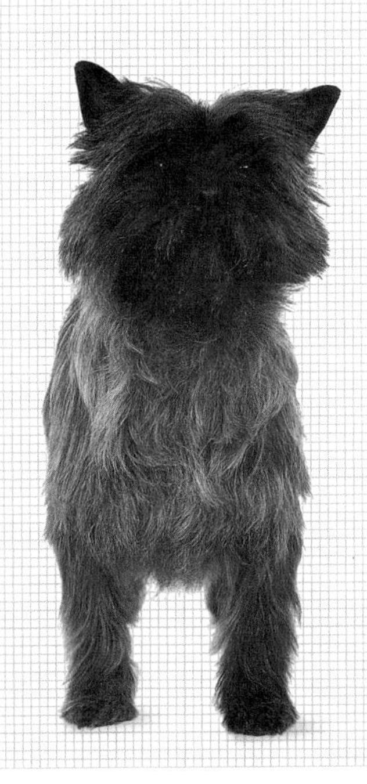

Specification

This breed is one of the smallest working terriers, with a rough, slightly unkempt appearance. Its double coat provides good protection against the elements, being coarse on top with a short, soft undercoat.

RECOGNITION Te
North America, Britain and FCI member countries

LIFESPAN
11–13 years

COLOUR
Any colour except white

HEAD
Broad, with a powerful, short muzzle

EYES
Well-spaced, protected by shaggy eyebrows

EARS
Widely separated, small and pointed; carried erect

CHEST
Moderate

TAIL
Held vertically

WEIGHT
Dogs 6.5kg (14 lb); bitches 6kg (13 lb)

CHILD FRIENDLINESS

GROOMING

FEEDING

EXERCISE

HEIGHT AT SHOULDER
Dogs 25.5cm (10 in); bitches 24cm (9½ in)

MALTESE

Possibly the oldest of all today's toy breeds, the Maltese has been highly valued since ancient times as a companion. It features prominently in writings and on artefacts from Classical times. These dogs are closely linked with the Mediterranean island of Malta, where they were probably introduced by the Phoenician traders. The modern type of the Maltese was established by the early 1800s, at a time when the dogs were actually becoming more scarce on Malta.

PERSONALITY

Do not be fooled by its small size – the Maltese is a fearless as well as a friendly breed. These dogs have highly affectionate and gentle natures, and they will happily spend much of the day in the company of their owner in the home.

HEALTH AND CARE

Problems related to abnormalities of the eyelashes can be an issue in this breed. The hairs rub against the surface of the eye, causing intense irritation and triggering excessive tear production. Watch for tear staining at the corners of the eye nearest the nose.

AS AN OWNER

Regular grooming is vital, although the lack of a dense undercoat makes the task quite easy. While grooming, check that the ears are clean – with such a lot of hair, they are susceptible to infection. The corners of the eyes may need to be wiped occasionally as well.

Specification

The long flowing coat lies flat and consists of a single layer, reaching almost to the ground around the sides of the body. The hair on the head is often tied up to form a characteristic topknot.

RECOGNITION T
North America, Britain and FCI member countries

LIFESPAN
11–13 years

COLOUR
White

HEAD
Well-balanced head, with a black nose

EYES
Dark brown and not prominent

EARS
Long, with good feathering, set low and lie on the sides of the head

CHEST
Balanced in proportion to the body

TAIL
Carried over the back, lying to one side over the hindquarters

WEIGHT
1.75–2.75kg (4–6 lb) preferred, but up to 3kg (7 lb) acceptable

CHILD FRIENDLINESS

GROOMING

FEEDING

EXERCISE

HEIGHT AT SHOULDER
25.5cm (10 in)

TIBETAN SPANIEL

The Tibetan Spaniel is not related in any way to other spaniels, but simply acquired this name in the West because of its resemblance to a toy spaniel. It is an ancient breed, traditionally linked with monastic communities in Tibet. Its ancestors used to turn prayer wheels, which carried the words of a prayer on a scroll, enabling these to be constantly repeated by the monks. Early Tibetan Spaniels were variable in size and appearance, with those near the Chinese border being more akin to the Pekingese (pp. 30–1) in their muzzle shape. The breed first reached the West in the 1890s.

PERSONALITY

An intelligent breed that relates well to people in its immediate circle, the Tibetan Spaniel is nevertheless suspicious of strangers. It also has an assertive side to its nature, but it is adaptable and does well in an urban environment.

HEALTH AND CARE

Tibetan Spaniels tend to be sounder than the Pekingese to whose ancestry they contributed, partly because their facial shape is less extreme. They are hardy dogs as well, thanks to their dense, double-layered coat. Their attractive appearance is easily maintained with standard grooming.

AS AN OWNER

Be aware that the coats of male Tibetan Spaniel puppies are likely to become more profuse than those of bitches, forming a small mane around the neck. Much of their dense undercoat is shed in the spring.

Specification

Today's Tibetan Spaniels have a much more standardized appearance, although they can still vary significantly in weight. The breed's forelegs are short and slightly bowed, supporting a relatively long body. The tail forms a plume, which is carried high over the body when the dog moves.

RECOGNITION NS
North America, Britain and FCI member countries

LIFESPAN
11–13 years

COLOUR
No restrictions

HEAD
Slightly domed skull, with a small head in proportion to the body

EYES
Medium-sized, dark, oval eyes

EARS
Medium-sized and set high

CHEST
Moderate in width

TAIL
Set high

WEIGHT
4–6.75kg (9–15 lb)

CHILD FRIENDLINESS

GROOMING

FEEDING

EXERCISE

NORWICH TERRIER

Originating from eastern England, the Norwich is something of an atypical terrier, not least because it traditionally worked in packs, rather than hunting independently. Its origins can be traced back to a livery stable in the city of Cambridge, with these terriers becoming popular with undergraduates there. One, called Rags, was given to another stable in the nearby city of Norwich in about 1900. Fourteen years later a direct descendant of Rags, called Willum, was taken to Philadelphia, where the breed was named after its British breeder, becoming known as the Jones Terrier. It gained show recognition in 1936.

PERSONALITY

Friendly, and with a more extrovert personality than many terriers, the Norwich makes an ideal companion dog, although it can display an assertive side to its nature on occasions.

HEALTH AND CARE

This is a tough, hardy breed. Its weatherproof coat needs little attention in terms of grooming.

AS AN OWNER

Be prepared for your dog to dig holes in your garden. Although the Norwich may represent an offshoot from the northern terrier lineage, it still displays this fairly characteristic terrier behaviour, particularly if left outdoors on its own for any length of time. Check that fences are on secure foundations, too, to ensure your terrier cannot burrow its way out of your garden.

Specification

Norwich Terriers have a lively appearance, thanks partly to their erect, pointed ears. The coat has a harsh, wiry texture and forms a ruff around the neck and shoulders, being shorter on the head, muzzle and ears. The front legs are short but strong, and the hindquarters muscular. The feet are rounded and the nails are black.

RECOGNITION Te
North America, Britain and FCI member countries

LIFESPAN
12–14 years

COLOUR
Red, wheaten, grizzle or black and tan

HEAD
Broad and rounded, with a powerful, relatively short muzzle

EYES
Small, oval and dark

EARS
Well-spaced, carried upright when alert

CHEST
Wide and deep

TAIL
Level with the topline and held erect

WEIGHT
5.5kg (12 lb)

CHILD FRIENDLINESS

GROOMING

FEEDING

EXERCISE

HEIGHT AT SHOULDER
25.5cm (10 in)

BORDER TERRIER

The smaller working terriers have much greater stamina than their size would suggest, and they are different in temperament from toy dogs of similar size. They are also more robust, as demonstrated by the Border Terrier. This breed was exhibited at agricultural events before it was seen at dog shows. Created in the Borders, as the area linking England and Scotland is known, this terrier has existed there for over 400 years. It was developed to run with the local hounds and yet to remain small and bold enough to venture underground and flush out a fox by itself.

PERSONALITY

Brave and fearless when working, the Border Terrier can be trained easily and is affectionate by nature. The breed has great stamina and is hardy as well.

HEALTH AND CARE

Border Terriers are sometimes born with holes in the walls of the ventricles of the heart, requiring surgical repair. In the case of mature male dogs, one or both testes sometimes fail to descend into the scrotum, potentially resulting in Sertoli cell tumours. Bitches occasionally suffer from primary uterine inertia, meaning that there are no contractions to push the puppies out at birth.

AS AN OWNER

Having been bred to work closely in the company of hounds, Border Terriers are reasonably well-tempered towards other dogs. In this they are unlike a number of other terrier breeds.

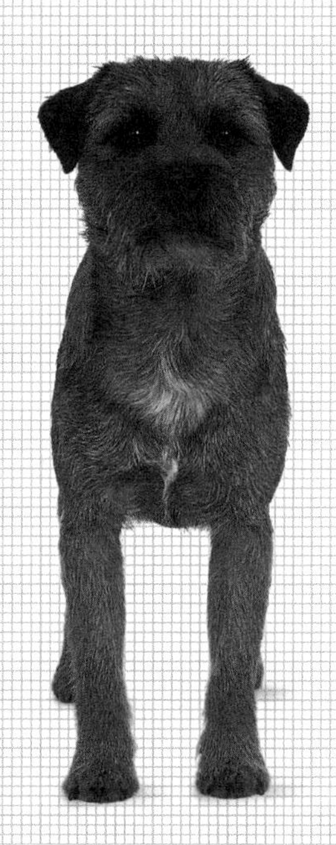

Specification

The Border Terrier is a relatively long-legged breed with a compact body shape. Its head has been likened to that of an otter. The topcoat is wiry, with a short dense undercoat, and the underlying skin is loose.

RECOGNITION Te
North America, Britain and FCI member countries

LIFESPAN
12–14 years

COLOUR
Blue and tan; wheaten; red, grizzle, and tan

HEAD
Broad, with a short muzzle

EYES
Medium-sized, dark hazel

EARS
Small and V-shaped, set on the sides of the face

CHEST
Neither deep nor narrow

TAIL
Short, tapered and not set high; held up when alert

WEIGHT
Dogs 6–7kg (13–15½ lb); bitches 5.25–6.5kg (11½–14 lb)

CHILD FRIENDLINESS

GROOMING

FEEDING

EXERCISE

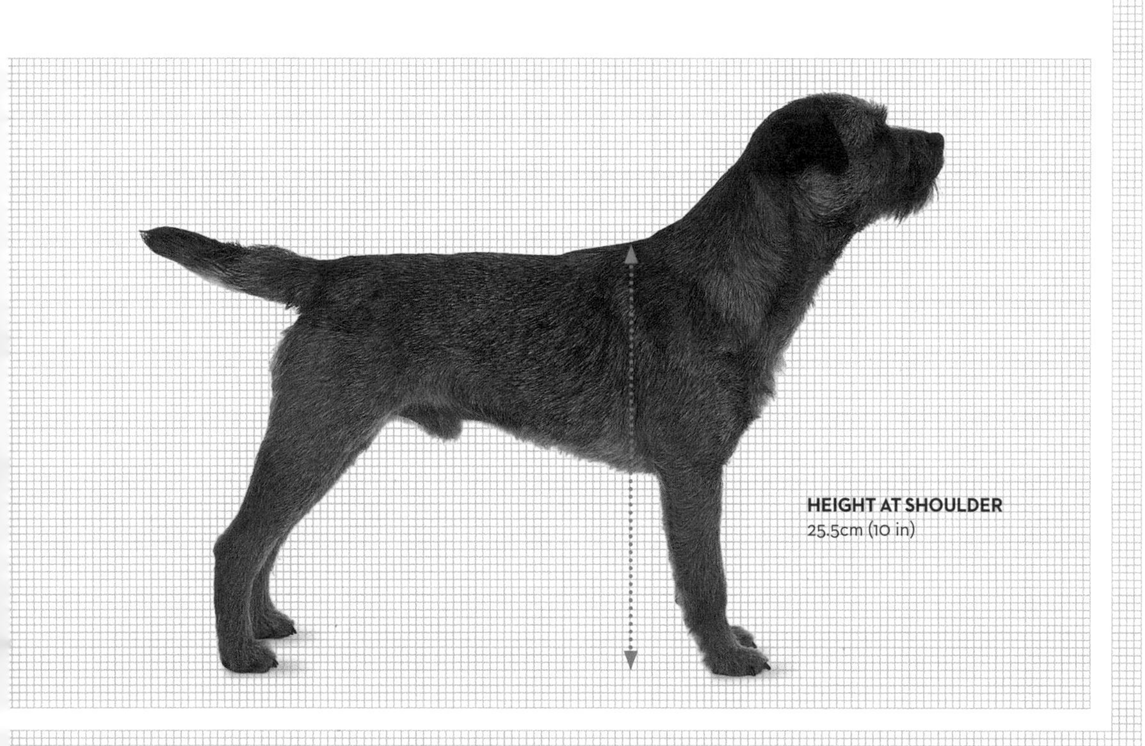

HEIGHT AT SHOULDER
25.5cm (10 in)

SCOTTISH TERRIER

Affectionately known as the Scottie, this terrier has an unmistakable appearance and a distinctive gait. Its back remains straight as it moves, and its legs are largely hidden by its trailing coat. This may be the oldest of all Britain's terrier breeds, dating back more than 500 years. Confusingly, it was originally called the Skye Terrier, but its name finally became established as the Scottish Terrier in 1879.

PERSONALITY

Determined and loyal, a Scottie makes an excellent companion for someone living alone. These are affectionate dogs, although not highly extrovert. The breed is famously headstrong and stubborn, and requires a strong-minded owner.

HEALTH AND CARE

Scotties may be afflicted by a genetic weakness known as Scottie cramp, resulting in short, painless seizures, lasting up to 30 seconds, during which time the dog's limbs, back and tail become rigid. The head may be held down between the front legs during a seizure, but recovery is rapid. Medical treatment can help, but will not cure this condition. The breed is vulnerable to skin cancer. Scotties also have a tendency to become deaf, so if your pet appears unresponsive, get its hearing checked.

AS AN OWNER

Try to socialize a Scottie as much as possible from an early age. Do not be surprised if a Scottie hunts small prey when off the leash, because the breed was originally bred as a ratter.

Specification

Scotties have a comparatively short body, powerful hindquarters and a tail that is carried erect. The coat is hard and wiry, with a soft undercoat.

RECOGNITION Te
North America, Britain and FCI member countries

LIFESPAN
12–14 years

COLOUR
Black, wheaten or brindle

HEAD
Long, with the square muzzle being of equal length

EYES
Dark, small and almond-shaped

EARS
Small, pricked and pointed, with the outer edges forming straight lines from the side of the skull

CHEST
Broad and deep

TAIL
Tapering, measuring about 18cm (7 in)

WEIGHT
Dogs 8.5–10kg (19–22 lb); bitches 8–9.5kg (18–21 lb)

CHILD FRIENDLINESS

GROOMING

FEEDING

EXERCISE

HEIGHT AT SHOULDER
25.5cm (10 in)

AUSTRALIAN TERRIER

This hardy little terrier is one of the smallest working breeds. Early in its history it acquired a remarkable reputation for catching venomous snakes, leaping up and seizing them behind the head. 'Aussies' will also hunt more typical terrier targets, including rabbits and rats. The breed had been created from assorted terrier breeds brought from Europe, ranging from the Dandie Dinmont (see pp. 28–9) to the Cairn (responsible for its red colouring, see pp. 42–3), by the 1870s. It is now represented in countries around the world, but it is still seen most commonly in its homeland.

PERSONALITY

Confident in spite of its small size, the Australian Terrier is naturally bold and inquisitive. It will form a strong bond with its owner and thrives on attention. This breed's senses are finely tuned, ensuring that it is effective as a ratter.

HEALTH AND CARE

Australian Terriers must be vaccinated against leptospirosis, a bacterial infection spread by rats, which can be fatal to dogs. The disease can be caught through contact with rats' urine.

AS AN OWNER

This breed is highly recommended for the proud home owner, because it does not shed, even in warmer climates and seasons – it also does not require much grooming. These terriers will settle down well in the home, whether as part of a family or with someone living alone, and make alert watchdogs.

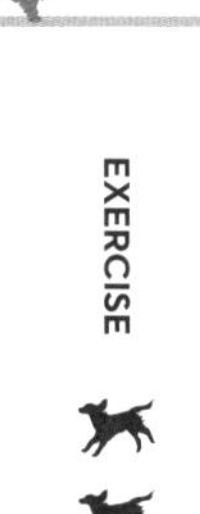

Specification

In profile, this terrier has a long body in proportion to its height. The outer coat has a harsh texture and is about 6.5cm (2½ in) long. The undercoat is soft and short, with a topknot of hair on the head.

RECOGNITION Te
North America, Britain and FCI member countries

LIFESPAN
12–14 years

COLOUR
Solid sandy or red; blue and tan

HEAD
Long and powerful; the skull length matches that of the strong muzzle

EYES
Well-spaced, small and dark

EARS
Small, pointed and held erect

CHEST
Extends below the elbows

TAIL
Set high and carried upright, almost vertically

WEIGHT
5.5–6.5kg (12–14 lb)

CHILD FRIENDLINESS

GROOMING

FEEDING

EXERCISE

HEIGHT AT SHOULDER
25.5–28cm (10–11 in)

WEST HIGHLAND WHITE TERRIER

This breed was created because of a tragic accident, when a Colonel Malcolm shot what he thought was a fox, only to find that he had killed his beloved Cairn Terrier (see pp. 42–3). He devoted his efforts then to creating an unmistakable white strain, to avoid confusion, using paler Cairn puppies. The emerging breed was originally called the Poltalloch Terrier, but ultimately became known as the West Highland White Terrier. These are typical strong-willed terriers, possessing personalities much larger than their size would suggest.

PERSONALITY

Bold and alert, the Westie makes a fearless guard dog in the home. These terriers are energetic, too, in spite of their size, but they may be combative with other dogs, especially of their own kind.

HEALTH AND CARE

Westies are susceptible skin complaints and various inherited conditions. Pain when eating, combined with a fever, is indicative of craniomandibular osteopathy, a juvenile problem that should resolve itself when the dog is about a year old. Stiffness in the pelvic region and a loss of coordination can, however, indicate the presence of progressive illness known as Krabbe's Disease.

AS AN OWNER

Professional grooming every two months, combined with regular brushing of the breed's hard top coat at home, should be adequate for a pet Westie. It can be difficult to keep the white coat pristine, because Westies enjoy outdoor games and energetic walks.

Specification

Westies are entirely white in colour. Their short front legs are strong and straight. The hind limbs are correspondingly short, with muscular thighs, while the back feet are smaller than the front ones.

RECOGNITION Te
North America, Britain and FCI member countries

LIFESPAN
11–13 years

COLOUR
White

HEAD
Broad head, rounded at the front, with a blunt muzzle

EYES
Medium-sized, almond-shaped and dark brown

EARS
Positioned on the outer edges of the head, and pointed

CHEST
Deep, reaching the elbows

TAIL
Short, carrot-shaped, straight and covered with hard hair; carried upright

WEIGHT
6.75–10kg (15–22 lb)

CHILD FRIENDLINESS

GROOMING

FEEDING

EXERCISE

HEIGHT AT SHOULDER
25.5–28cm (10–11 in)

COTON DE TULEAR

The Coton de Tulear's origins lie on the small Indian Ocean island of Reunion. French settlers there brought small Bichon-type dogs with them during the 1700s, which interbred with the local canine population, ultimately creating this breed. Individuals originating in Reunion were traded at the port of Tulear (now Toliara) on Madagascar (explaining the second part of its name). It became a fashionable and exclusive pet among wealthy families on this bigger island, but remained unknown elsewhere until the mid-20th century, being introduced to Europe, and then to North America in 1974.

PERSONALITY

Bred simply as a companion, the Coton de Tulear is ideally suited to domestic living. It has a placid, friendly temperament, combined with a playful side to its nature.

HEALTH AND CARE

The breed is generally healthy, but there is only a small gene pool, which could see problems arising in future. Regular claw trimming is often necessary, to prevent the nails from becoming overgrown.

AS AN OWNER

The Coton de Tulear is not as widely-kept as some toy breeds, so you may have to be patient and be prepared to travel to obtain a puppy.

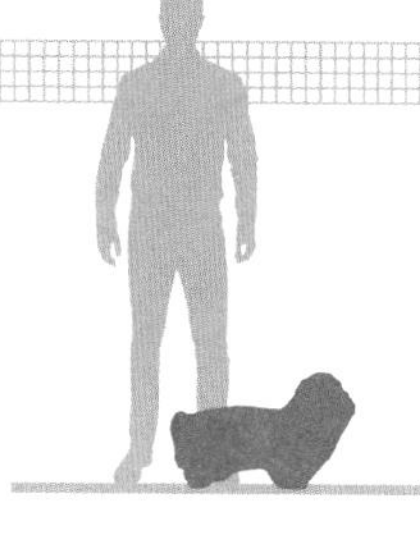

Specification

Its white fluffy coat has a texture like cotton (as reflected by the French word 'coton') with a soft, light feel. Although white is favoured, individuals with cream, or even black, patches are not unknown. The Coton de Tulear has a profuse covering of hair on the head, extending over the eyes. Both the nose and the claws are black.

RECOGNITION T
FCI member countries

LIFESPAN
10–12 years

COLOUR
Pure white, bicolour (white and cream/white and black) or tricolour (white, cream and beige)

HEAD
Skull is longer than the muzzle, with the head in proportion to the body

EYES
Medium-sized, relatively round and dark

EARS
Narrow and hang closely down the sides of the head

CHEST
Deep

TAIL
Carried raised; curled loosely over the back, with only the hair touching here

WEIGHT
5.5–7kg (12–15 lb)

CHILD FRIENDLINESS

GROOMING

FEEDING

EXERCISE

HEIGHT AT SHOULDER
25.5–30.5cm (10–12 in)

SCHIPPERKE

This breed used to be seen on barges plying their trade along the canals of Flanders, acting as a watchdog. The name of the Schipperke should be pronounced 'sheep-er-ker', almost without sounding the final 'r'. It is seemingly descended from spitz-type stock, but it could be a scaled-down form of the now extinct Belgian breed of sheepdog called the Leauvenaar. Schipperkes took part in probably the first-ever dog show back in 1690, at the Grand Palace of Brussels, wearing distinctive custom-made hammered brass collars, which traditionally they still wear today. The breed was first seen in North America in 1888.

PERSONALITY

Lively, with an innate curiosity, the Schipperke also acts as an alert guardian, barking without hesitation at the approach of strangers. Outdoors, these dogs will also often display terrier-like ratting skills.

HEALTH AND CARE

The Schipperke is a hardy breed, with a reputation for longevity. Its dense undercoat provides good insulation, and even if the coat becomes wet, water runs off readily, rather than penetrating down to the skin. For show purposes, the quality of the coat is vital, and there must be clear differentiation between the outer and undercoat.

AS AN OWNER

Energetic by nature, the Schipperke is a good choice for a home alongside children, often forming a particularly strong bond with them.

Specification

These dogs are black in colour, often with a slightly paler undercoat, and become grey around the muzzle as they get older. This breed is often naturally tail-less, a feature that emphasizes its square profile.

RECOGNITION NS
North America, Britain and FCI member countries

LIFESPAN
13–16 years

COLOUR
Black

HEAD
Medium-width with a tapering muzzle

EYES
Dark brown, small, oval-shaped

EARS
Small and triangular; set high and held erect

CHEST
Deep and broad

TAIL
When present, carried below the level of the body

WEIGHT
5.5–7kg (12–16 lb)

CHILD FRIENDLINESS

GROOMING

FEEDING

EXERCISE

HEIGHT AT SHOULDER
Dogs 28–33cm (11–13 in); bitches 25.5–30.5cm (10–12 in)

PEMBROKE WELSH CORGI

The name 'Corgi' probably derives from the Celtic words for 'watch' and 'dog', with records of similar dogs existing as far back as the 10th century. The Pembroke Corgi has a more fox-like appearance than its Cardigan relative (see pp. 72–3), thanks to its smaller ears, and originated in the old southern Welsh county of Pembrokeshire. When the Pembroke and Cardigan Corgis were recognized as individual breeds in 1934, the Pembroke was by far the more numerous. It remains more popular, being the favoured breed of Queen Elizabeth II who kept Pembroke Welsh Corgis from 1933 until 2018, and was recognized in North America back in 1937.

PERSONALITY

Pembroke Corgis are just as inclined to nip. This is instinctive rather than aggressive behaviour.

HEALTH AND CARE

Pembroke Welsh Corgis have a hereditary predisposition to a weakness of the intervertebral discs in the neck, so always exercise these corgis using a harness rather than a collar, to protect the neck from injury. They can also suffer from epilepsy.

AS AN OWNER

Your corgi may sometimes nip your ankles in impatience when anticipating a walk or a meal. They can also be quarrelsome with other dogs, so bear this in mind when exercising your pet. Corgis also suffer from more birthing difficulties than many breeds.

Specification

Unlike the Cardigan, the Pembroke Welsh Corgi is typically shorter in the back, as well as smaller and lighter in weight. The colour range is more restricted, with no blue merle form. The short legs often turn slightly inward.

RECOGNITION He
North America, Britain and FCI member countries

LIFESPAN
11–13 years

COLOUR
Sable, fawn, red and black-and-tan, sometimes with white areas

HEAD
Muzzle slightly tapered

EYES
Oval, medium-sized

EARS
Medium-sized, erect and tapered slightly to rounded points

CHEST
Deep

TAIL
Short

WEIGHT
Dogs 12.25–13.5kg (27–30 lb); bitches 11.5–12.5kg (25–28 lb)

GROOMING

FEEDING

EXERCISE

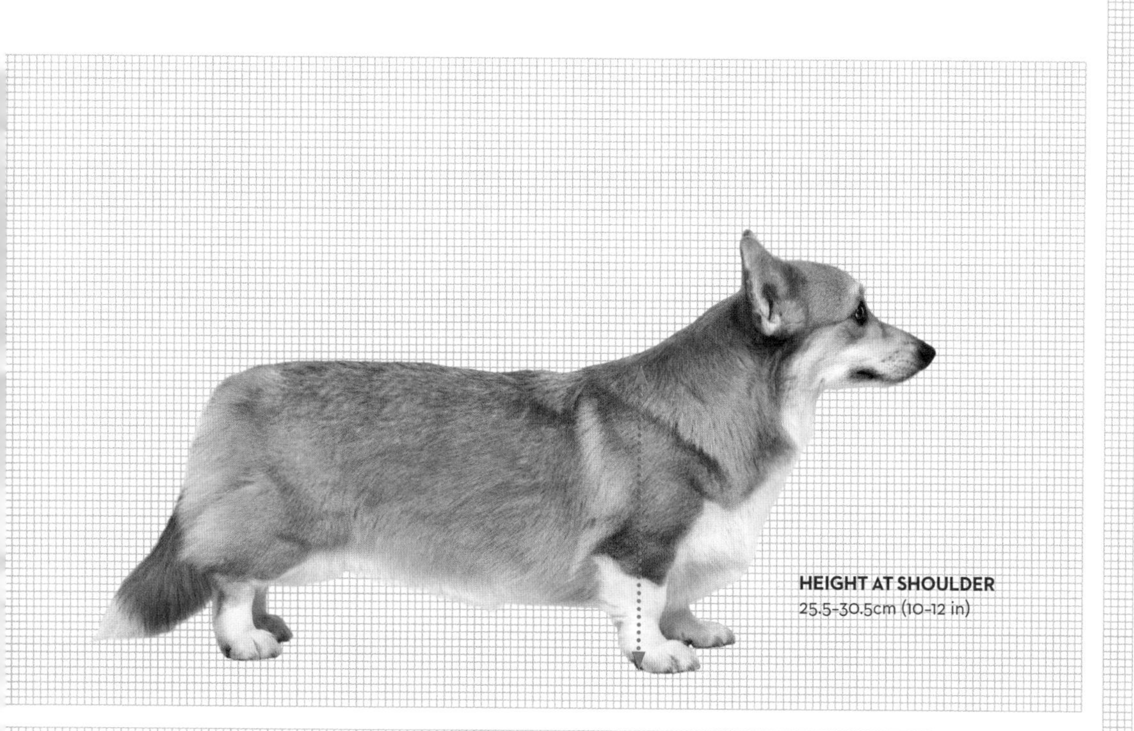

POMERANIAN

The Pomeranian's spitz ancestry is clearly shown by its erect ears, foxlike face and a tail that curls forwards over the back. But while larger spitz breeds pull sleds, the Pom has always been a companion breed. Named after the German province of Pomerania, this breed was created in Britain in the 1800s by down-sizing the Mittel Spitz through selective breeding, and it became a favourite of Queen Victoria.

PERSONALITY

Lively and energetic, the Pomeranian retains attributes seen in its larger relatives, but it settles well as a household pet, even where space is limited. Alert by nature, a Pom is an excellent watchdog.

HEALTH AND CARE

Unfortunately, Pomeranians are susceptible various congenital problems found in small dogs. These include patellar luxation, where affected puppies typically develop signs of lameness at 4–6 months. This weakness can cause dislocation of the kneecap, affecting one or both hind legs, and is likely to require surgical correction. Subsequently, dogs generally recover well.

AS AN OWNER

Be prepared to groom your pet daily to prevent the coat from becoming matted. Wipe the corners of the eyes with damp cotton wool if tearstaining is evident, as is most likely during cold weather.

Specification

The ancestor of the Pomeranian was white, but today the breed has more than a dozen colour variations. The Pomeranian is sometimes described as the Puffball, because of its thick, erect coat, which arises from the combination of a soft, dense undercoat and a long, straight, harsh-textured topcoat. There is a distinct frill on the chest, extending around to the shoulders on each side of the body.

RECOGNITION T
North America, Britain and FCI member countries

LIFESPAN
11-13 years

COLOUR
No restrictions

HEAD
Muzzle quite narrow, creating a fox-like appearance

EYES
Medium-sized and dark

EARS
Small and erect, set quite high

CHEST
Relatively deep, but not too wide

TAIL
Carried flat and straight over the back

WEIGHT
Dogs 1.75-2kg (4-4½ lb); bitches 2-2.5kg (4½-5½ lb)

CHILD FRIENDLINESS

GROOMING

FEEDING

EXERCISE

HEIGHT AT SHOULDER
28cm (11 in)

PUG

Pugs probably originated in China nearly 2,000 years ago. They then became popular in the Netherlands, before being introduced to Britain in the late 1700s. The breed was originally fawn, with a black mask and saddle markings, and the popular black Pug was unknown until the 1870s. Their original name of Pug Dog means 'monkey-faced dog', and arose because of their resemblance to marmoset monkeys.

PERSONALITY

Pugs make excellent pets for owners of all ages. They are tolerant and patient, and will not yap to gain attention, although they make effective watchdogs and have been known to defend their owners against attack.

HEALTH AND CARE

The Pug's prominent eyes are prone to injury, and they sometimes have ingrowing eyelashes, which inflame the eyeballs. The wrinkled skin on the face may be a focus for infection, although this is not common. Very little grooming is required. Avoid exercising a Pug when the weather is very hot, to protect against heatstroke.

AS AN OWNER

Unfortunately, this breed is susceptible to obesity, so carefully measure out the amount of food required for your Pug and only offer healthy treats, such as pieces of raw carrot or apple, with regular periods of exercise also being essential.

Specification

One of the biggest toy breeds, the Pug is recognizable by its stocky build and thick, powerful legs. Its face is flat and wrinkled, with the eyes being large and prominent. The ears hang down over the sides of the head and the tail curls forwards low over the back. The Pug's body has a square-shaped profile and its fur has a fine, soft texture.

RECOGNITION T
North America, Britain and FCI member countries

LIFESPAN
13–15 years

COLOUR
Silver or apricot-fawn with black markings; solid black

HEAD
Massive and round, with a short square muzzle

EYES
Prominent, dark and large

EARS
Thin, small, and either the preferred button shape or roselike

CHEST
Broad

TAIL
Twisted tightly, with a double curl desirable

WEIGHT
6.5–8kg (14–18 lb)

GROOMING

FEEDING

EXERCISE

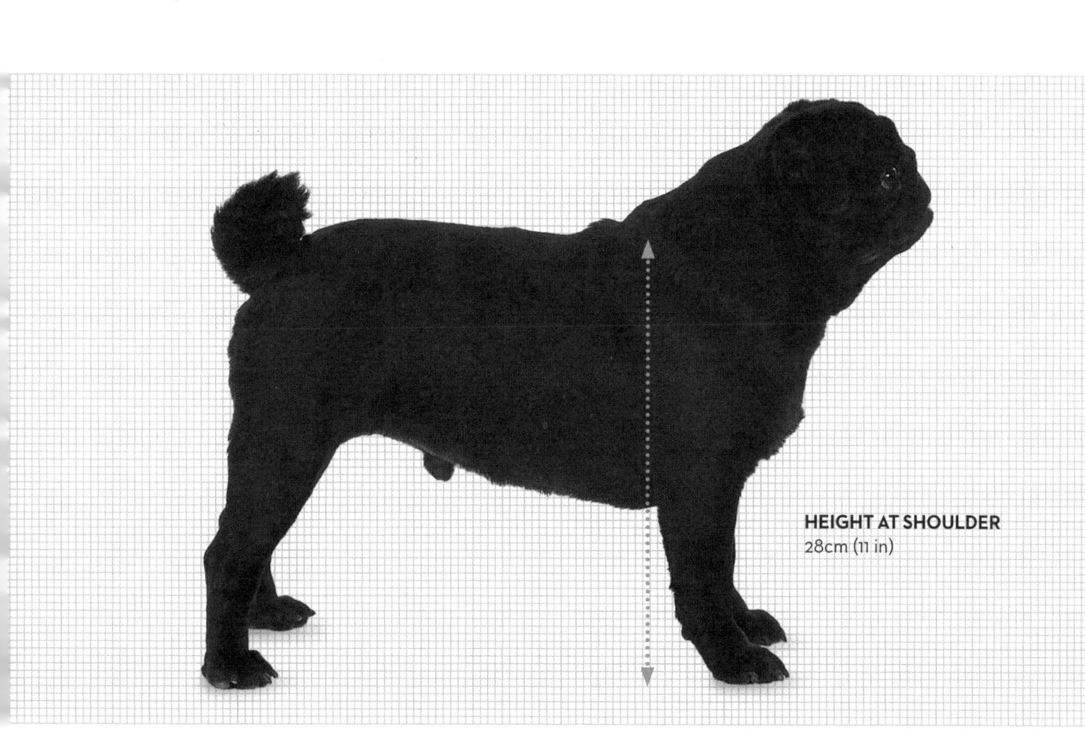

PAPILLON

The Papillon has inspired many famous artists including Titian, Goya, Rubens and Rembrandt. It originated in mainland Europe over 500 years ago, being bred in the Italian city of Bologna. It was then taken by mule to the French court of Louis XIV, and became a highly fashionable lap dog, in addition to being an effective ratter. Remarkably though, it remained unknown in Britain and North America until the early 1900s.

PERSONALITY

The breed is friendly and highly affectionate. It has peaceful and adaptable disposition, being just as happy trotting gracefully in open fields as it is in an urban park.

HEALTH AND CARE

Despite appearing delicate, these continental toy spaniels are hardy but, as with other toy dogs, they can suffer from a weakness that afflicts the kneecaps. Grooming is easy, in spite of the long coat, because the Papillon does not have a thick undercoat.

AS AN OWNER

A Papillon can adjust to changes in your life without problems, being ideal both for people living on their own and for those with families. Unsurprisingly, this breed learns quickly and is very responsive.

Specification

'Papillon' in French means 'butterfly', and the name refers to the large raised ears of the breed, which are said to look like a butterfly's spread wings. They are set slightly to the sides of the head and are fringed with hair. It is important that the coloured areas of fur must not only cover the ears entirely, but also extend from the ears to the eyes.

RECOGNITION T
North America, Britain and FCI member countries

LIFESPAN
11–13 years

COLOUR
Chestnut and white; red and white; black and white; tricolour-black, tan and white

HEAD
Slightly rounded on top, with a delicately pointed muzzle

EYES
Round, dark, and set low on the skull

EARS
Large, erect, located far back, and with rounded tips

CHEST
Relatively deep, separating straight, fine-boned front legs

TAIL
Long and carried high over the back, creating a plume

WEIGHT
4.5kg (10 lb)

CHILD FRIENDLINESS

GROOMING

FEEDING

EXERCISE

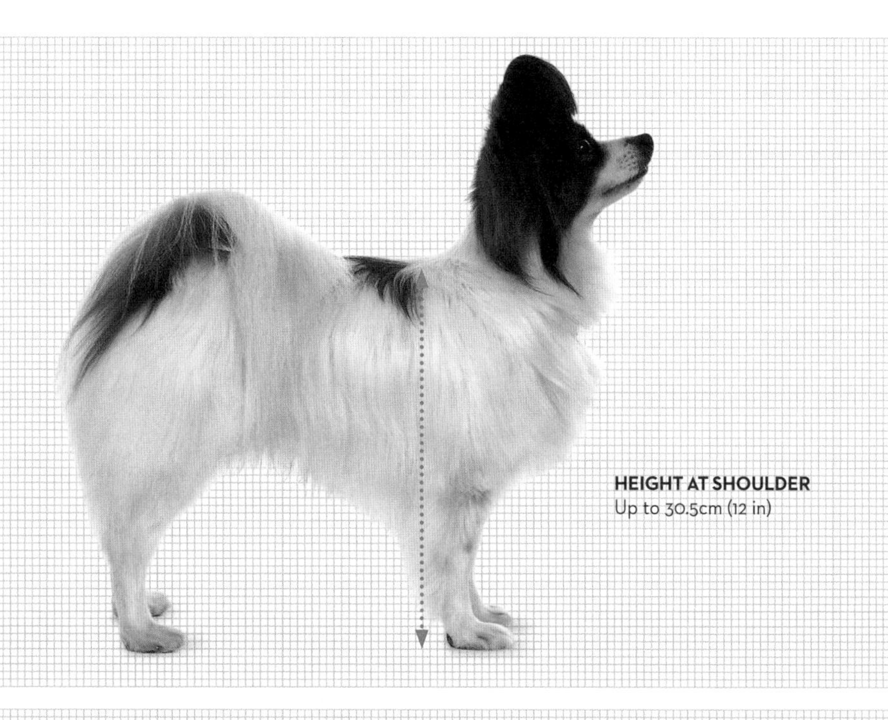

HEIGHT AT SHOULDER
Up to 30.5cm (12 in)

MINIATURE PINSCHER

Also known as the MinPin, this breed has a distinctive high-stepping gait, resembling that of a hackney horse. The Miniature Pinscher literally prances around the show ring, holding a front leg out straight and then bending it at the joint before putting it down again. In common with other Pinschers, it hails from Germany, originating there in the 1700s and being known as the Zwergpinscher. The MinPin was originally kept to hunt vermin, and the designation 'pinscher' is now synonymous with 'terrier'.

PERSONALITY

This breed has no concept of its small size – the MinPin is bold by nature and individuals are usually willing not only to stand their ground, but also to challenge much larger dogs if threatened. It also excels as a guard dog.

HEALTH AND CARE

An inherited breed weakness that may be seen occasionally in the Miniature Pinscher is a tendency to dislocate its shoulder. If this occurs, you need to seek veterinary advice to prevent any recurrence in the future, and this may entail surgery. Otherwise, in spite of its delicate appearance, it is a robust breed.

AS AN OWNER

Train your pet carefully to avoid potential conflicts when you are out walking and to ensure that your dog returns to you readily on command. Intelligent by nature, this is a breed that learns quickly. Coat care is straightforward: occasional brushing will maintain the breed's sleek and attractive appearance.

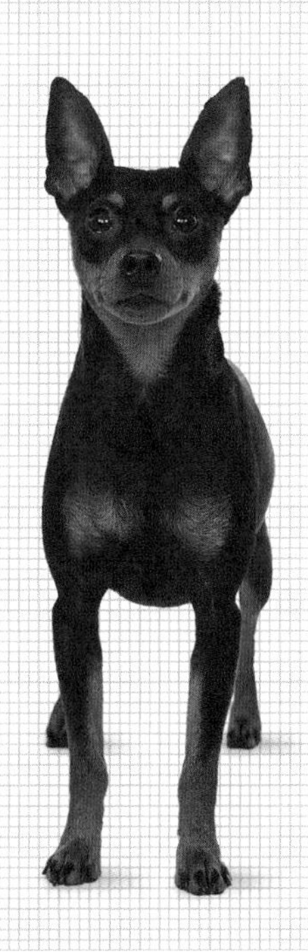

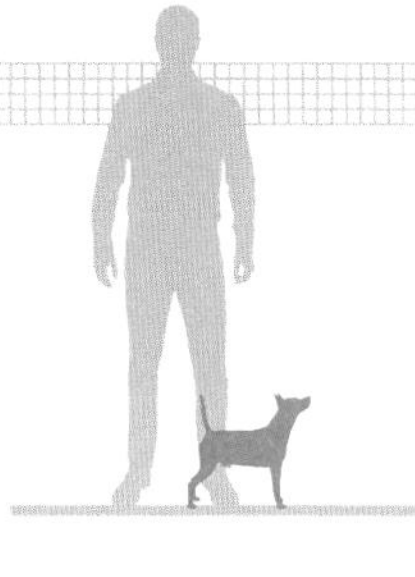

Specification

A well-muscled athletic profile, emphasized by its short glossy coat. The breed used to be larger in size than it is today. The feet are catlike in shape, with strong pads and broad, blunt claws.

RECOGNITION T
North America, Britain and FCI member countries

LIFESPAN
11–13 years

COLOUR
Solid red; chocolate and red; black and red

HEAD
Slightly elongated in appearance

EYES
Slightly oval, dark, almost round

EARS
Small in size, set high and erect

CHEST
Full and relatively broad

TAIL
Set relatively high, continuing the top line

WEIGHT
5.5kg (12 lb)

CHILD FRIENDLINESS

GROOMING

FEEDING

EXERCISE

HEIGHT AT SHOULDER
25.5–31.5cm (10–12½ in)

CARDIGAN WELSH CORGI

The Cardigan breed is named after the old Welsh county where it was created, separating it from its better-known relative, the Pembroke Corgi (see pp. 62–3). Their origins in Wales may date back over a millennium. Their ancestry is mysterious, although these breeds may somehow be related to the Swedish Vallhund, which has a similar appearance. The short legs of corgis make them effective as cattle drovers – they will encourage a reluctant cow to keep moving by nipping at its heels, with little risk of being kicked.

PERSONALITY

Tough and hardy, these corgis have a surprisingly dominant personality, plus a very energetic nature. They definitely should not be considered as lap dogs.

HEALTH AND CARE

Cardigan Welsh Corgis can suffer from the hereditary eye condition called progressive retinal atrophy (PRA), which ultimately leads to blindness, so breeding stock should have been screened. The dense undercoat is shed in spring, when your dog will need a good deal of grooming.

AS AN OWNER

These working farm dogs retain their instinctive desire to nip, and this must be curbed as much as possible from puppyhood. It makes them unsuited for a home with young children.

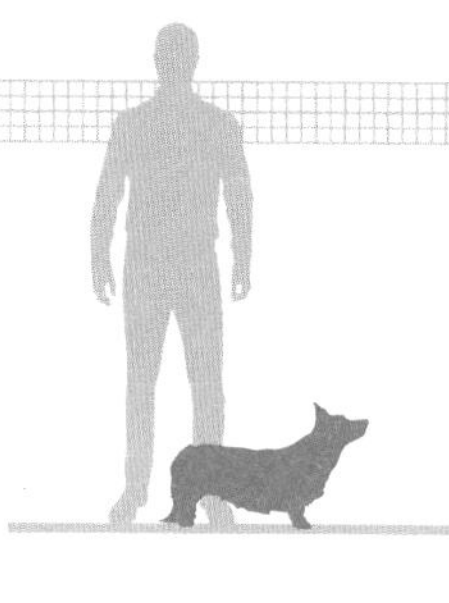

Specification

The Cardigan is larger and longer than its tail-less Pembroke relative. It is known in Welsh as 'Ci Llathaid', literally 'yard long', indicating its length from the top of its nose to the tip of its tail. The Cardigan's ears are also much larger and more rounded than those of the Pembroke.

RECOGNITION He
North America, Britain and FCI member countries

LIFESPAN
11–13 years

COLOUR
Sable, brindle and red; rarer blue merle or black dogs may have tan and/or brindle points

HEAD
Flat top with the muzzle being shorter than the skull

EYES
Medium to large and widely spaced

EARS
Large and prominent, with rounded tips

CHEST
Evident breastbone; medium-width with deep brisket

TAIL
Set and carried low, never above the back

WEIGHT
11.5–15.5kg (25–34 lb)

GROOMING

FEEDING

EXERCISE

HEIGHT AT SHOULDER
26.5–31.5cm (10½–12½ in)

CHINESE CRESTED

The distinctive Chinese Crested exists in two forms, both of which may occur in the same litter. There can be Hairless puppies, or those with a full coat called Powderpuffs. The breed achieved a lot of publicity in the USA during the late 1800s because the famous burlesque dancer Gypsy Rose Lee chose to keep this near-naked dog!

PERSONALITY

A playful nature, combined with an attractive personality, win over even those who instinctively dislike the idea of a hairless dog. The Chinese Crested is lively by nature, with an agile, trotting gait.

HEALTH AND CARE

The hairless gene in dogs is usually linked to a reduction in the number of teeth, with premolar teeth lacking in the Hairless Chinese Crested, although Powderpuffs usually have a full complement of teeth. The lack of fur in Hairless examples means they generally do not attract fleas, and their skin naturally feels warm to the touch.

AS AN OWNER

Hairless dogs need protection both from both heat and cold. A coat or special sweater – or both – will be required in cold or wet weather. On sunny days, apply a special canine sunblock to protect against sunburn, which can lead to skin cancer, especially on unpigmented areas of skin.

Specification

Although described as 'hairless', hair is evident on the Chinese Crested's head and tail, as well as 'socks' of hair on the lower legs and feet. The hair on the head is called the crest, and that on the tail is described as the plume. In contrast, the Powderpuff variety has a true double coat, which is long, with a silky texture.

RECOGNITION T
North America, Britain and FCI member countries

LIFESPAN
10–12 years

COLOUR
No restrictions

HEAD
Balanced head shape

EYES
Widely spaced and almond-shaped

EARS
Large and erect, set at the same level as the eyes

CHEST
Deep brisket, but breastbone is not prominent

TAIL
Slim and tapered, forming a curve and reaching to the hock; carried erect

WEIGHT
up to 5.5kg (12 lb)

CHILD FRIENDLINESS

GROOMING

FEEDING

EXERCISE

HEIGHT AT SHOULDER
28–33cm (11–13 in)

CAVALIER KING CHARLES SPANIEL

This spaniel is one of the most popular toy breeds, being closely related to the King Charles Spaniel, but distinguishable by its much longer facial shape and smaller size. Both trace their ancestries back to the court of King Charles II of England in the second half of the 17th century. Over time however, selective breeding had created spaniels with an increasingly flattened face.

PERSONALITY

Cavaliers have a friendly, well-adjusted nature. They are not athletic compared with larger spaniels, but they will display a playful side to their natures, especially if encouraged to take part in games from an early age.

HEALTH AND CARE

These spaniels are greedy and prone to obesity. This, in turn, predisposes them to diabetes, to which the breed is susceptible. Their flowing, silky coats must have regular grooming. Some Cavaliers have a strange nervous condition that causes them to jump up occasionally, as though they were trying to catch nonexistent flies.

AS AN OWNER

Do not rely on food treats when training this breed, because of its tendency to put on weight – use healthy options, such as small pieces of carrot or apple. Apart from the risk of diabetes, obesity may worsen heart problems, to which these spaniels are also susceptible, especially in later life.

Specification

The Cavalier King Charles Spaniel has been developed into a slightly taller breed than the King Charles Spaniel, also known as the English Toy Spaniel. It is characterized by its significantly longer muzzle, meaning that its facial appearance overall is less compact. The colour varieties in both breeds are identical.

RECOGNITION T
North America, Britain and FCI member countries

LIFESPAN
10–12 years

COLOUR
Chestnut and white; black and tan; red; tricolour

HEAD
Relatively flat top to the head, with a tapering muzzle

EYES
Round, large and dark

EARS
Set high, long and hang down the sides of the head

CHEST
Medium-sized

TAIL
Balanced relative to the body, generally carried low

WEIGHT
5.5–8kg (12–18 lb)

CHILD FRIENDLINESS

GROOMING

FEEDING

EXERCISE

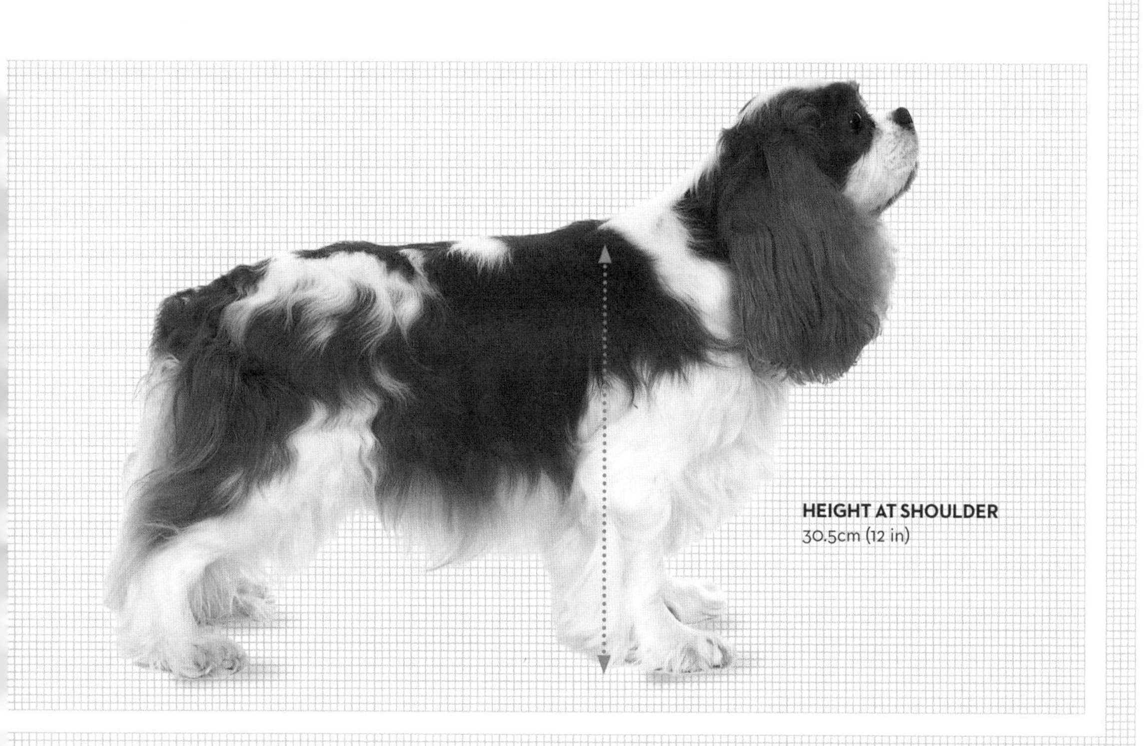

HEIGHT AT SHOULDER
30.5cm (12 in)

FRENCH BULLDOG

These popular small dogs are easily recognized by their characteristic batlike ears. During the mid-19th century, a Toy Bulldog breed was developed in the industrial towns of England. It was taken to northern France by lacemakers when they emigrated there, with the breed then being further developed by crossings with terriers that resulted in its characteristic pricked ears. The French Bulldog gradually became a fashionable pet in Paris, and soon gained an international following, notably in North America.

PERSONALITY

Active, friendly and easily trained, the French Bulldog makes an excellent companion dog. Known affectionately as a 'Frenchie', this dog makes an alert guard dog, but is not noisy or excitable by nature.

HEALTH AND CARE

French Bulldogs are sometimes born with a malformation of the spinal vertebrae, which is not always immediately apparent. This breed occasionally develops haemophilia, an inherited blood disorder. These dogs can also be prone to flatulence. Their grooming needs are minimal.

AS AN OWNER

Makes a personable pet, whether living in a family with children or with a single owner. Like other breeds with large square heads, French Bulldogs sometimes have difficulty in giving birth naturally and may need a Caesarean birth.

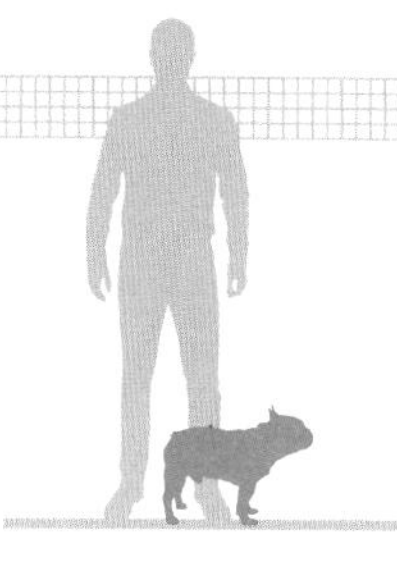

Specification

The top of the skull is flat and the forehead slightly rounded. The raised ears are positioned at a high point on the head. The chest is broad and, despite the breed's small size, the French Bulldog is well-muscled with straight forelegs. The tail is short and may have a slightly screwed appearance, but it is not actually curly.

RECOGNITION
North America, Britain and FCI member countries

LIFESPAN
10–12 years

COLOUR
Most colours but not solid black

HEAD
Skull flat between the ears, with a deep broad muzzle

EYES
Widely spaced, round and medium-sized, set away from the ears

EARS
Distinctive bat-eared appearance, set high; broad at the base, with round tips

CHEST
Deep and broad

TAIL
Set low, short and tapered

WEIGHT
Up to 12.5kg (28 lb)

CHILD FRIENDLINESS

GROOMING

FEEDING

EXERCISE

HEIGHT AT SHOULDER
30.5cm (12 in)

MINIATURE SCHNAUZER

Bred originally to hunt rats, but with a different heritage from that of other members of the terrier group, the Miniature Schnauzer will not venture underground to catch its quarry. The process of miniaturization leading to its development began with the selective breeding of the smallest puppies from Schnauzer litters (see pp. 130–1). In the late 1800s, Miniature Schnauzers were then crossed with Affenpinschers to scale down the emerging breed even further.

PERSONALITY

Lively, playful and alert, the Miniature Schnauzer is a breed that can be trained with little difficulty. It relates well to children and, unlike some terriers, it is usually tolerant of other dogs.

HEALTH AND CARE

This breed is generally sound, in spite of being scaled down. Puppies can be afflicted by a congenital narrowing of the pulmonary artery, causing breathlessness and lack of energy, and the problem requires surgical correction if possible. This weakness was probably inherited from the Schnauzer itself.

AS AN OWNER

Coat care is demanding because it requires careful stripping to maintain its distinctive appearance. If a Miniature Schnauzer is simply clipped, the characteristic banding in the top coat will be lost and the undercoat will show through, and this can add to the cost of ownership, with professional grooming assistance required. Having a bark that sounds like that of a much larger dog, it makes a good guardian.

Specification

The unusual colour combinations seen in the Miniature Schnauzer result from alternating light and dark banding running down the individual hairs. The hindquarters of this breed are especially angular, and the hock joints of the legs can extend further behind the body than the tail.

RECOGNITION Te
North America, Britain and FCI member countries

LIFESPAN
12–14 years

COLOUR
Salt and pepper, black and silver, black

HEAD
Rectangular, with a strong, blunt muzzle

EYES
Dark brown and small

EARS
Set high, small, V-shaped, and naturally folded on the skull

CHEST
Moderate

TAIL
Set high and carried upright

WEIGHT
6–6.75kg (13–15 lb)

CHILD FRIENDLINESS

GROOMING

FEEDING

EXERCISE

HEIGHT AT SHOULDER
30.5–35.5cm (12–14 in)

BULLDOG

This breed has changed dramatically in appearance, as it was formerly much taller, with illustrations from a century ago suggesting that Bulldogs were then more like Boxers (see pp. 174–5). They were originally bred to take part in bull-baiting contests, leaping up onto the bull's head, and clinging on with their strong jaws. When bull-baiting was banned in 1835, the Bulldog became one of the first breeds to be recognized for show purposes, shrinking significantly in stature.

PERSONALITY

Today's Bulldog is a placid breed, with a dignified bearing. It is stoic, too, as its appearance suggests, and it is a loyal and affectionate pet. This dog does not require a great deal of space, and will get along well with others.

HEALTH AND CARE

Bulldogs have various congenital problems, often affecting the heart and spine, so check the breeding stock carefully before acquiring a puppy and take your dog to the vet soon after bringing it home. Bitches frequently have difficulty giving birth and require an emergency Caesarean section, because of the size of the puppies' head. The coat is short, smooth and glossy, and needs little grooming.

AS AN OWNER

Look out for eye irritations and any signs of infection in the folds of facial skin. Avoid exercising a Bulldog in hot weather; the breed is prone to heatstroke, with potentially serious consequences if your dog has a weak heart.

Specification

A broad, massive head with an upturned nose and strong jaws are features recalling the breed's past. The legs are powerful, short and bowed, with the shoulders being powerful. Bulldogs are now popular companions, with a distinctive appearance and a characteristic rolling gait.

RECOGNITION NS
North America, Britain and FCI member countries

LIFESPAN
8–10 years

COLOUR
Brindle, piebald, white, red, fawn and fallow

HEAD
Exceedingly large, with a short, upturned muzzle

EYES
Set low and rounded

EARS
Rose-eared, positioned high on the skull

CHEST
Deep and broad

TAIL
Short, tapers and hangs low

WEIGHT
Dogs 22.5kg (50 lb); bitches 18kg (40 lb)

CHILD FRIENDLINESS

GROOMING

FEEDING

EXERCISE

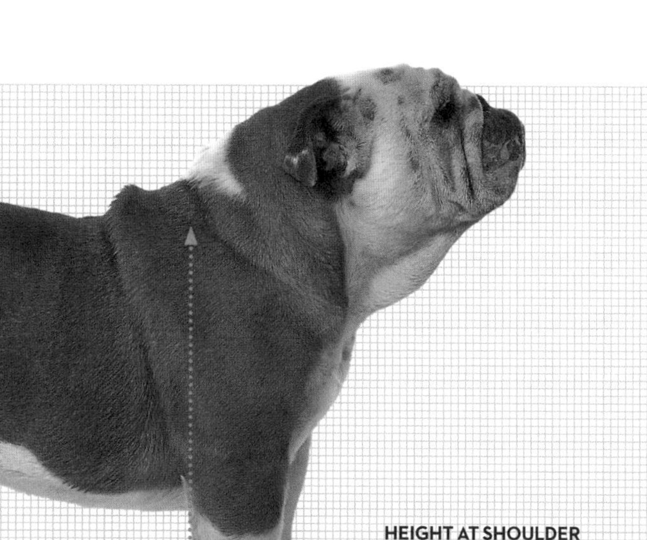

HEIGHT AT SHOULDER
30.5–35.5cm (12–14 in)

GLEN OF IMAAL TERRIER

This ancient breed has only come to international prominence since the late 20th century, achieving full American Kennel Club registration in 2004, although the Irish Kennel Club recognized it back in 1933. It has changed remarkably little over the 400 years of its existence, being named after its place of origin in County Wicklow, Ireland. It is descended from native Irish breeds, perhaps crossed with a basset of some type. The Glen of Imaal Terrier has undertaken roles as varied as that of a turnspit dog (turning the wheel connecting to the spit roasting meat over a kitchen fire), fighting dog and hunter of badgers. Despite this varied past, this breed has always been a genial companion dog.

PERSONALITY

Unlike many terriers, the Glen of Imaal has a quiet nature. It can be surprisingly athletic and runs fast, like a hound. It settles down readily as a family pet.

HEALTH AND CARE

These terriers are generally fit and hardy. The natural deviation in the position of their front feet means that their nails may wear unevenly, and these will need to be trimmed back.

AS AN OWNER

Good socialization, as with other terriers, is important, and be prepared to pay for professional grooming, for hand-stripping of the coat and nail care.

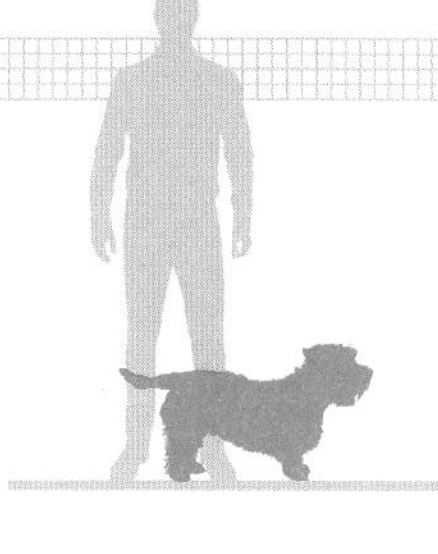

Specification

Long-bodied and low-slung, the Glen of Imaal Terrier has a harsh, weather-resistant top-coat, with a softer undercoat beneath. It is often wheaten, but in a wide range of shades from cream through to a reddish tone. The front feet turn naturally outward.

RECOGNITION Te
North America, Britain and FCI member countries

LIFESPAN
12–14 years

COLOUR
Wheaten, blue or brindle

HEAD
Strong, with a powerful muzzle

EYES
Well-spaced, round, brown and medium-sized

EARS
Small and widely spaced; rose or half-pricked when alert

CHEST
Strong, wide and deep, reaching below the elbows

TAIL
Strong and carried above the level of the back

WEIGHT
16kg (35 lb)

CHILD FRIENDLINESS

GROOMING

FEEDING

EXERCISE

HEIGHT AT SHOULDER
31.5–35.5cm (12½–14 in)

PARSON RUSSELL TERRIER

Parson Russell Terriers were traditionally bred for their working abilities, and there was no serious attempt to standardize their appearance. They rank as the most popular English terrier, proving to be personable companions. The breed's name commemorates that of its founder, Parson Jack Russell, who lived in the southwest of England during the mid-1800s. He used Fox Terriers (see pp. 104–5 and 108–9) in its development, and they were originally bred to dig foxes out from underground burrows.

PERSONALITY

Bold, lively and excitable, this is a breed that retains strong working instincts. The Parson Russell is not particularly patient and may express frustration by nipping, which means that it is not an ideal choice for a home with young children.

HEALTH AND CARE

Today's standard reflects that of the original strain, which was bred to be athletic, and capable of running easily alongside longer-legged Foxhounds. These terriers often prove to be skilled escape artists, good both at digging and at jumping surprisingly high, so double-check your garden is suitably secure.

AS AN OWNER

Parson Russell terriers will readily 'go to ground', investigating tunnels and rabbit holes. When out in the countryside, watch your pet to ensure that you know its whereabouts, because there is otherwise a risk that it may get into trouble. Retrieving a trapped terrier from underground is difficult.

Specification

The Parson Russell Terrier is noticeably taller than many ordinary Jack Russell Terriers, but is otherwise similar. The coat can be either broken (rough) or smooth. The breed's small chest enables it to go underground easily.

RECOGNITION Te
North America, Britain and FCI member countries

LIFESPAN
12–14 years

COLOUR
White and black, tan and white, or tricolour

HEAD
Flat skull with a powerful, rectangular muzzle

EYES
Dark and medium-sized

EARS
V-shaped, small, and folded near the top of the head

CHEST
Narrow, yet quite deep, creating an athletic appearance

TAIL
Not set high, creating a level topline; carried high

WEIGHT
6–7.75kg (13–17 lb)

GROOMING

FEEDING

EXERCISE

HEIGHT AT SHOULDER
Dogs 35.5cm (14 in); bitches 33cm (13 in)

BEAGLE

The Beagle makes an energetic, exuberant companion, with a playful side to its nature. Possessing considerable stamina, it delights in long walks through the countryside. Beagles were originally bred to hunt hares in packs, accompanied by hunt followers on foot. Their origins are now unknown, but date back over 400 years. Beagles first started to become popular in the United States during the 1860s, and usually get on well with other dogs.

PERSONALITY

Lively and instinctively friendly, Beagles make excellent companions, although they will often tend to disappear off on the trail of a scent rather than wait on command or return to you. In other respects, they make an ideal choice as a pet.

HEALTH AND CARE

Beagles can easily become seriously overweight and suffer from associated complications ranging from diabetes to heart trouble. Some Beagle strains are susceptible to epilepsy, but fits rarely occur within the first year of life. Careful veterinary management and medication should stabilize this condition.

AS AN OWNER

Do not be misled by their small size – Beagles need plenty of exercise, including the opportunity to run off the leash every day. If bored and underexercised, a Beagle is likely to be destructive around the home. They are very affectionate, forming strong bonds with members of the family. Like other hounds, they will not thrive in crowded urban city areas where exercise options are limited.

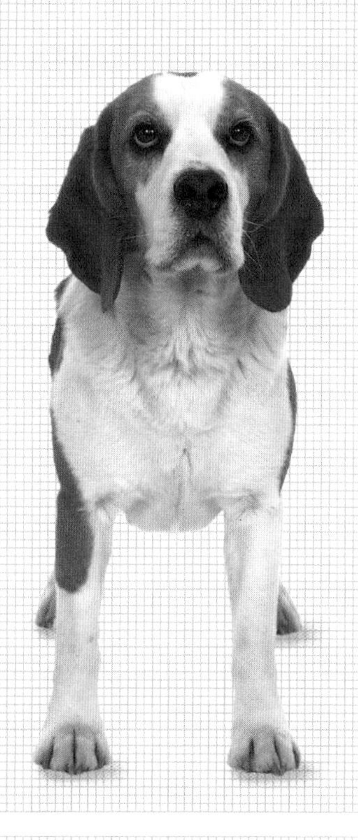

Specification

This breed has straight legs and appealing brown eyes. Beagles occur in the same colours as the Basset Hound (pp. 98–9), again displaying highly individual markings. The coat itself has a hard texture and is short, making it easily groomed.

RECOGNITION H
North America, Britain
and FCI member countries

LIFESPAN
10–12 years

COLOUR
Typically bi- or tricoloured

HEAD
Relatively long
and slightly domed

EYES
Well-space and large

EARS
Set relatively low a
nd broad in shape

CHEST
Deep and broad

TAIL
Set quite high, but short and
carried with a slight curve

WEIGHT
8–13.5kg (18–30 lb)

CHILD FRIENDLINESS

GROOMING

FEEDING

EXERCISE

HEIGHT AT SHOULDER
(two categories): up to 33cm
(13 in) and 33–38cm (13–15 in)

ITALIAN GREYHOUND

This toy breed may represent the earliest example of selective breeding in the dog world – being created as a miniature version of the Greyhound (see pp. 232–3), and then kept exclusively as a companion rather than as a hunting breed. The mummified remains of similar dogs have been discovered in the tombs of the ancient Egyptian pharaohs. Later the breed was a popular sight at the royal courts of Europe, as a lady's companion.

PERSONALITY

This breed can be shy and nervous with strangers, but socialization from an early age should help to overcome problems. Italian Greyhounds are gentle, trustworthy dogs and make excellent companions.

HEALTH AND CARE

As often happens with deliberate miniaturization, the soundness of the Italian Greyhound was compromised by the late 1800s. Now, thanks to careful breeding, these small hounds are again robust. Italian Greyhounds prefer to run fast on flat ground, so try to find some open country to exercise your dog.

AS AN OWNER

Italian Greyhounds feel the cold and dislike the wet. They should wear coats for protection in bad weather. They need very minimal grooming – they have no undercoat, but the lack of insulation makes them especially sensitive to the elements. On the move, Italian Greyhounds have a distinctive, high-stepping gait not seen in their larger relatives.

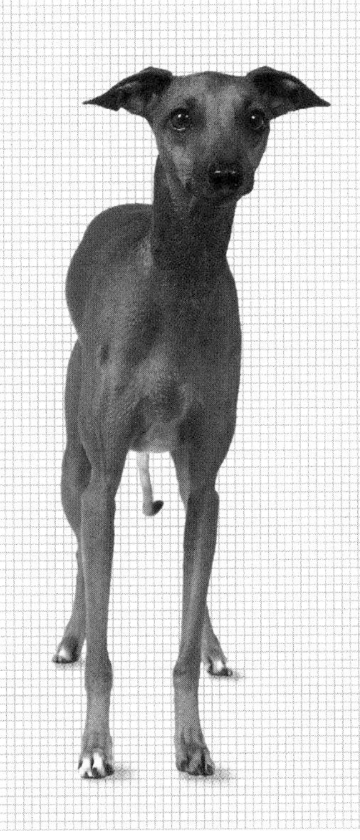

Specification

As with its larger relative, the Italian Greyhound has the typical deep chest and arched, sloping back that assist it to sprint at speed on its slender legs. The coat is thin, short and slightly glossy. There may often be white markings present on the chest, or forming a blaze between the eyes.

RECOGNITION T
North America, Britain and FCI member countries

LIFESPAN
11–13 years

COLOUR
Most colours accepted

HEAD
Flat and long; narrow muzzle

EYES
Bright and relatively large

EARS
Soft, set back on the head and rose-shaped

CHEST
Narrow and deep

TAIL
Fine, relatively long, and carried low

WEIGHT
3.5kg (8 lb)

GROOMING

FEEDING

EXERCISE

HEIGHT AT SHOULDER
33–38cm (13–15 in)

SHETLAND SHEEPDOG

These sheepdogs, often described as Shelties, originate from the Shetland Islands, off Scotland's northwest coast. Their origins are a mystery but they are similar to the Rough Collie (albeit smaller; see pp. 184–5), which originated on the mainland.

PERSONALITY

The Sheltie has a loyal, affectionate and friendly nature towards people it knows well, but it is also an alert guard dog and does not accept strangers readily. Few breeds are as quick to learn.

HEALTH AND CARE

This breed is prone to hip dysplasia, arising from a congenital malformation of the hip joints. Shelties are also susceptible to a number of eye problems, including progressive retinal atrophy (PRA), so appropriate checks need to be carried out as far as possible before acquiring a puppy.

AS AN OWNER

Thorough, regular grooming is essential to maintain this dog's attractive appearance and prevent its coat from becoming matted. This is an agile breed and can jump well, which means that your garden must be adequately fenced. If you want a dog with which to enter junior handling or agility competitions, the Sheltie is an excellent choice.

Specification

The hair on the face is short, and the head is framed by a ruff behind the ears, which is at its most profuse during the winter. Sable colouring can vary from a golden shade to mahogany. Some Shelties may have an unusual blue and white iris (the coloured area around the pupil), which is known as a 'wall eye'.

RECOGNITION He
North America, Britain and FCI member countries

LIFESPAN
11–14 years

COLOUR
Sable, blue merle, or black with white and/or tan markings

HEAD
Blunt, wedge-shaped and tapered from ears to nose

EYES
Medium-sized, black and white iris and set obliquely

EARS
Stand largely erect when alert; folded lengthways at rest

CHEST
Deep, with brisket reaching to the point of the elbows

TAIL
Hangs straight down or slightly curved at rest; otherwise raised, but never curved over the back

WEIGHT
6.5–7kg (14–16 lb)

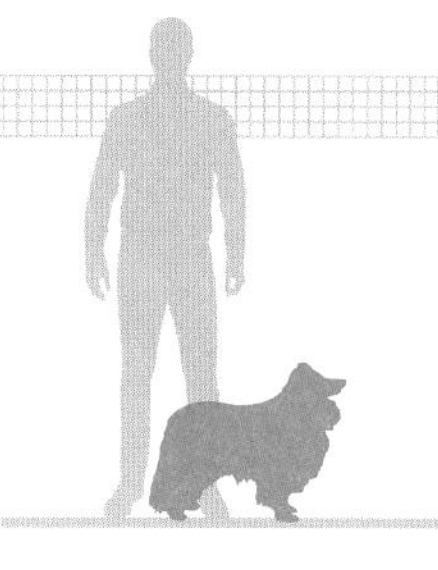

CHILD FRIENDLINESS

GROOMING

FEEDING

EXERCISE

HEIGHT AT SHOULDER
33–40.5cm (13–16 in)

TIBETAN TERRIER

In spite of its name, the Tibetan Terrier was bred to herd sheep rather than for hunting vermin, and still carries out this role in its homeland today. It deserves to be more popular as a pet, simply because it makes such a wonderful companion. The breed was unknown in the West until 1895, and little is known about its ancestry. Its long coat is clipped in the summer with those of the sheep, and used together with yak hair to make cloth.

PERSONALITY

This breed is affectionate, sensitive and responsive. Its loyalty is illustrated by one story in which a Tibetan Terrier saved its owner by driving off a rabid dog, and suffered a fatal bite as a consequence.

HEALTH AND CARE

Both sound health and temperament have long been regarded as important characteristics of this breed in Tibet. Remarkably, Tibetan Terriers do not seem particularly bothered by climatic extremes, whether hot or cold, and this is probably a reflection of the climate in their homeland.

AS AN OWNER

Be prepared for puppies having much shorter, single-layered coats, which are also likely to have a softer texture than those of adults.

Specification

The Tibetan Terrier has been described as a miniature Old English Sheepdog, which gives some impression of its appearance. The breed is well-equipped to survive outdoors in the bitter winters of its homeland, thanks to its profuse, double-layered coat. The undercoat is woolly, which helps to trap warm air close to the body, and the outer coat is fine and, in some cases, wavy.

RECOGNITION NS
North America, Britain and FCI member countries

LIFESPAN
11–13 years

COLOUR
No restrictions at all

HEAD
Medium-length

EYES
Widely spaced, large and dark

EARS
V-shaped, hanging down loosely over the sides of the head

CHEST
Moderate in width

TAIL
Set high and curled over the back; well-feathered

WEIGHT
9–13.5kg (20–30 lb)

GROOMING

FEEDING

EXERCISE

HEIGHT AT SHOULDER
Dogs 35.5–40.5cm (14–16 in); bitches 33–35.5cm (13–14 in)

SHIBA INU

This breed is the most common native Japanese breed in its homeland, sometimes also known as the Brushwood Dog, based on the terrain where it is traditionally used to hunt game birds. Today, however, it is now popular simply as a companion. The ancestors of the Shiba Inu have been present in Japan for over 2,500 years. They developed their unmistakable spitz appearance following crosses with imported dogs around AD 200, and since then the breed has changed remarkably little in appearance.

PERSONALITY

Highly alert and intelligent, the Shiba Inu makes a good companion, which helps to explain its rapidly increasing international popularity. It is a breed that has a long history of working alongside people, and it is responsive as a result.

HEALTH AND CARE

The breed is generally healthy, but as with all dogs, it is important to keep their vaccinations up to date. The Shibu Inu may perhaps be more susceptible than many dogs to the distemper virus. This infection almost wiped out the breed in its homeland after the end of World War II in 1945.

AS AN OWNER

Protective by nature, the Shiba Inu makes a good guarddog. It has an unusual bark, that sounds more like a shriek than a conventional bark. Effective training is necessary as these dogs still often retain keen hunting instincts.

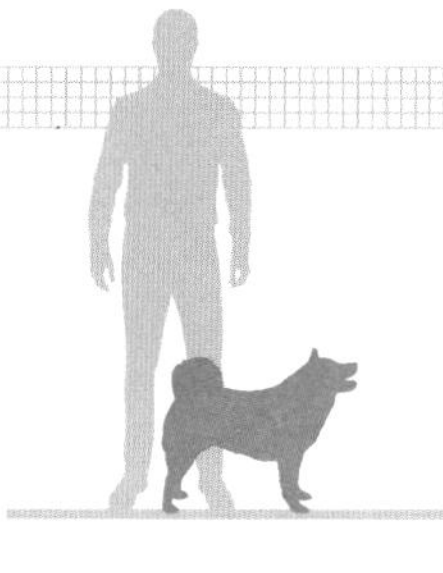

Specification

The Shiba Inu is the smallest of all Japan's native breeds; its name actually means 'small dog'. It has an alert appearance, thanks to its ears, which tilt forwards, and a kindly expression. The coat is short, with a harsh topcoat above a soft, dense undercoat, providing good insulation. The hair is longest on the curved tail.

RECOGNITION NS
North America, Britain and FCI member countries

LIFESPAN
11–13 years

COLOUR
Red, black, and sesame, which has a red undercoat and black tips to the fur, with a paler undercoat and markings

HEAD
Skull has a broad forehead, with a round muzzle

EYES
Deep-set and dark

EARS
Well-spaced, pricked, and triangular

CHEST
Deep

TAIL
Carried upward, over the back, either curled or sickle-shaped

WEIGHT
Dogs 10.5kg (23 lb); bitches 7.75kg (17 lb)

CHILD FRIENDLINESS

GROOMING

FEEDING

EXERCISE

HEIGHT AT SHOULDER
Dogs 37–42cm (14½–16½ in); bitches 34–39cm (13½–15½ in)

BASSET HOUND

Various hounds are described as 'bassets', with the Basset Hound being the best-known example. The name comes from the French 'bas', meaning 'low', referring to their short legs. Unusually, however, there is not a taller form of this particular breed. The ancestors of today's Basset Hounds were brought to England from France in 1866. Crossbreeding with Bloodhounds (see pp. 196–7) served to improve the Basset Hound's scenting skills, as well as resulting in a stockier build and a longer head.

PERSONALITY

A friendly, exuberant breed, the Basset Hound has a rich, baying call, frequently heard when it scents its quarry. The breed is not especially easy to train, but has a very affectionate nature.

HEALTH AND CARE

Puppies are vulnerable to inguinal hernias, where part of the abdomen is displaced. The third vertebra in the neck is sometimes malformed at birth, causing pressure on the spinal cord, and needs surgical correction, as will ectropion, when the eyelids hang away from the eyes. Very little grooming is needed, thanks to the breed's short coat.

AS AN OWNER

Be aware that these hounds may head off if they discover a scent when you are out walking. Their naturally social nature means that they will live together in relative harmony if you want more than one dog. Basset Hounds are prone to obesity, being gluttonous by nature.

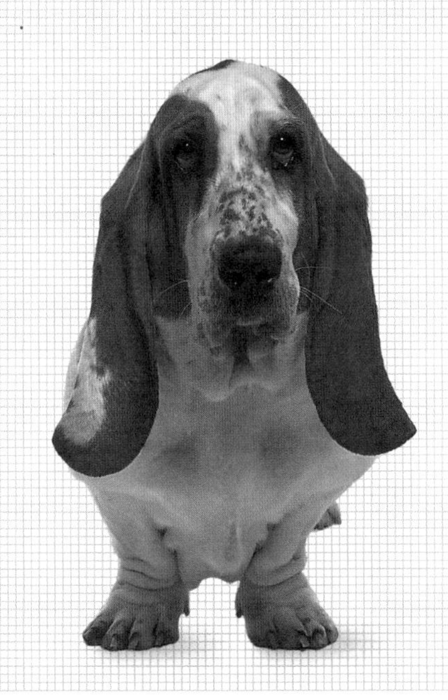

Specification

A domed head and long pendulous ears are characteristic of this breed, with the skin on the legs being markedly wrinkled in appearance. Their feet are remarkably large in comparison with their legs. They have long tails, which ensure that they can be seen even in deep undergrowth.

RECOGNITION H
North America, Britain and FCI member countries

LIFESPAN
10–12 years

COLOUR
Any hound colour – often bi- or tricoloured

HEAD
Large, with a well-domed skull

EYES
Slightly sunken, showing the haws, and brown in colour with a sad expression

EARS
Long, set low and far back on the head

CHEST
Long, broad rib cage

TAIL
Curves upwards

WEIGHT
22.5kg (50 lb)

CHILD FRIENDLINESS

GROOMING

FEEDING

EXERCISE

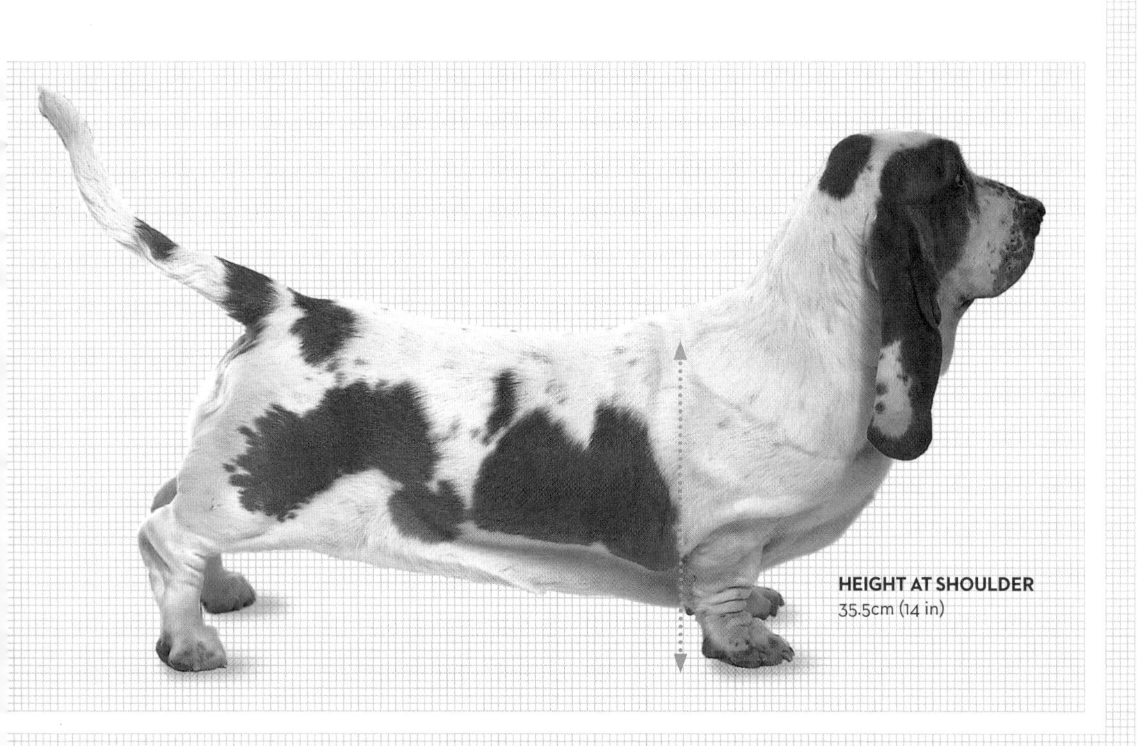

HEIGHT AT SHOULDER
35.5cm (14 in)

AMERICAN COCKER SPANIEL

During the 1870s, English Cocker Spaniels (see pp. 116–7) became more common in North America, being used primarily to hunt game birds, and they gradually developed into a quicker, small version of their ancestor. Known in the United States simply as the Cocker Spaniel, it is now the smallest spaniel bred originally for sporting purposes, being capable of both flushing game from brush and retrieving on land and in water.

PERSONALITY

Easy-going and enthusiastic, the American Cocker Spaniel is a tireless and talented sporting companion. It is affectionate, forms a close bond with those in its immediate circle and settles easily in the home.

HEALTH AND CARE

Unfortunately, the soundness and temperament of the American Cocker Spaniel have suffered because of its popularity. At least 25 different hereditary and congenital health problems have been recognized in this case, although puppies from a reputable breeder will come from stock that has been screened for many of them. It is still strongly advised that you have your new pet checked by your vet.

AS AN OWNER

Daily grooming is essential, with particular attention being paid to hair on the long pendulous ears, which are susceptible to infections. Seek veterinary help promptly if your dog starts scratching its ears repeatedly.

Specification

Apart from its smaller size and lighter weight, the most obvious difference between the American Cocker Spaniel and its English relative is its much longer coat. When seen in profile, its back is shorter, and the head is more domed with the muzzle being blunter too.

RECOGNITION S
North America, Britain and FCI member countries

LIFESPAN
10–12 years

COLOUR
Any solid colour, may have tan points, and parti-coloured, including roans

HEAD
Rounded, with a deep broad muzzle and square jaws

EYES
Round, pointing directly forwards

EARS
Set low; long and well-feathered

CHEST
Deep, extending down the elbows

TAIL
Carries on with the top line of the back

WEIGHT
11–12.5kg (24–28 lb)

CHILD FRIENDLINESS

GROOMING

FEEDING

EXERCISE

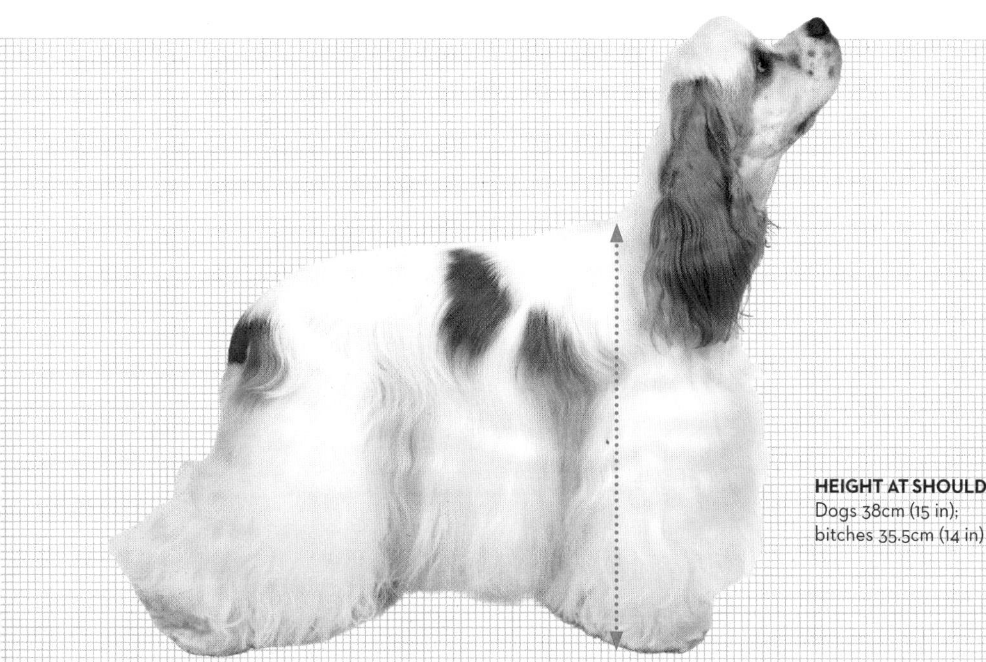

HEIGHT AT SHOULDER
Dogs 38cm (15 in); bitches 35.5cm (14 in)

STAFFORDSHIRE BULL TERRIER

Originally bred from fighting stock, Staffies can be problematic with other dogs, although they frequently make excellent companions for people. They shares their origins with the Bull Terrier, with both breeds originally resulting from matings between Bulldogs and Black and Tan Terriers. Whereas the head shape of the Bull Terrier altered dramatically with further crosses, that of this breed stayed the same, with broad jaws suggestive of mastiff stock. To distinguish between them, this breed was given the name of the English county where it was especially popular when it was first recognized in 1935.

PERSONALITY

Committed, intelligent and tenacious, Staffies are loyal and frequently playful at home with their families, but may become argumentative when out in a park where there are other dogs.

HEALTH AND CARE

Male Staffies should always be neutered to lessen the risk of aggressive behaviour. There is a breed disposition to cataracts as the result of a recessive gene. Grooming needs are minimal.

AS AN OWNER

Ensure as always that you are in charge of your dog at all times, particularly when out walking together. If your pet is aggressive with other dogs, you can fit it with a muzzle, or exercise it when it is unlikely to meet other dogs. Staffordshire Bull Terriers have a tendency to become fat if they are overfed and under-exercised, so do watch your pet's weight.

Specification

Smooth-coated and with a thick set and an extremely muscular appearance, the Staffordshire Bull Terrier has strong straight front legs, although the obvious strength of these dogs is also evident in their thick hind legs.

RECOGNITION Te
North America, Britain and FCI member countries

LIFESPAN
11–13 years

COLOUR
Fawn, red, black or blue, brindle, plus white markings or white

HEAD
Short, broad head plus powerful, broad muzzle

EYES
Round and medium-sized

EARS
Small; rose or half-pricked

CHEST
Broad

TAIL
Set and carried low; medium-length

WEIGHT
Dogs 12.5–17kg (28–38 lb); bitches 11–15.5kg (24–34 lb)

CHILD FRIENDLINESS

GROOMING

FEEDING

EXERCISE

HEIGHT AT SHOULDER
35.5–40.5cm (14–16 in)

WIRE FOX TERRIER

Early on, Wire and Smooth Fox Terriers (see pp. 108–9) were frequently bred together, although today they are treated as separate breeds. They are very similar in temperament, however, and are quite demanding choices as pets. Although the Wire has become closely related to the Smooth Fox Terrier, the rough-coated Black and Tan Terriers played a greater part in its development. Thanks to repeated cross-breeding, these Fox Terriers now essentially differ only in their coat type, with the Wire Fox Terrier being more commonly seen.

PERSONALITY

An alert but independent-minded breed, the Wire Fox Terrier can be wilful on occasions. These terriers make excellent guardians though, but they can also be noisy, barking at the slightest sound.

HEALTH AND CARE

The same congenital defects, including a higher-than-usual incidence of deafness occur in both types of Fox Terrier. Be alert to any swelling in the neck region, which may be indicative of the thyroid gland disorder described as goitre.

AS AN OWNER

Wire Fox Terriers are very enthusiastic about digging, and require a securely fenced garden. They are also keen and able ratters, being both fast and agile. Professional trimming of the coat will be needed to maintain their attractive appearance.

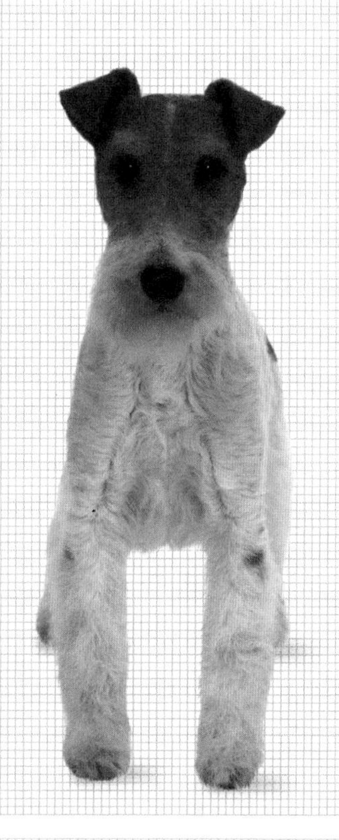

Specification

Its outer coat is dense and feels wiry, with a soft undercoat beneath. It should be longest over the back and sides of the body. The hairs may twist but should not be curly. The length of the head should be between 18 and 18.5cm (7–7¼ in) in a dog, and is slightly shorter in a bitch.

RECOGNITION Te
North America, Britain and FCI member countries

LIFESPAN
11–13 years

COLOUR
White, plus areas of tan or black.

HEAD
Almost flat top to the head and a large muzzle

EYES
Small and not widely spaced

EARS
V-shaped, small and set high

CHEST
Deep, yet not too broad

TAIL
Strong, set high and held upright

WEIGHT
7.25–8kg (16–18 lb)

GROOMING

FEEDING

EXERCISE

HEIGHT AT SHOULDER
Dogs 39cm (15½ in); bitches 37cm (14½ in)

ENGLISH TOY TERRIER

The Manchester Terrier exists in two sizes, with the smaller form often described as the English Toy Terrier. They used to be exhibited together but are now grouped separately. The hometown of these terriers is the city of Manchester, where they became popular during the 19th century. A fashion to breed smaller versions of the Manchester Terrier grew. These small dogs were nevertheless skilled ratters, and ratting contests with associated gambling became common. The most famous ratter of this era was called Tiny the Wonder, who became renowned for killing 300 rats in just three hours in 1848, despite weighing only 2.5kg (5 lb).

PERSONALITY

Lively and loyal, and although the Toy form of the Manchester Terrier has a true terrier personality, yet it usually gets along well with other dogs, even if they are much larger.

HEALTH AND CARE

The trend towards miniaturization led to an overall breed weakness – the natural consequence of persistently breeding weaker, smaller dogs from Manchester Terrier litters together – and so the Toy form had become scarce by the early 1900s. Since then, however, great efforts have been made to ensure the breed is generally strong and healthy. When individual dogs have health problems today, these are most likely to be skin and eye conditions.

AS AN OWNER

A dog that needs very little grooming, and will prove to be a big character.

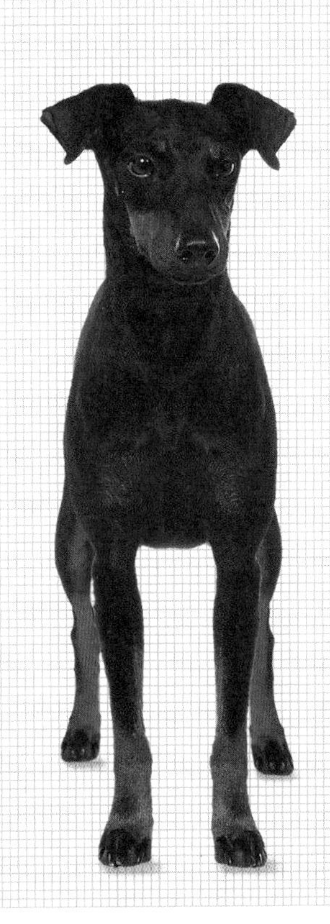

Specification

In the Toy, only naturally erect ears – described as 'candle-flamed' – are acceptable, whereas in the Standard, they may be folded down – creating what is known as a button ear.

RECOGNITION T
North America, Britain and FCI member countries

LIFESPAN
10–12 years

COLOUR
Clearly-defined areas of black and tan

HEAD
Long and narrow

EYES
Almond-shaped, virtually black, and bright

EARS
Wide at the base, tapering to points and set high; can be erect or button-shaped

CHEST
Brisket is close to the elbows

TAIL
Tapered to a point; carried in a slightly upward curve

WEIGHT
4.5–10kg (12–22 lb)

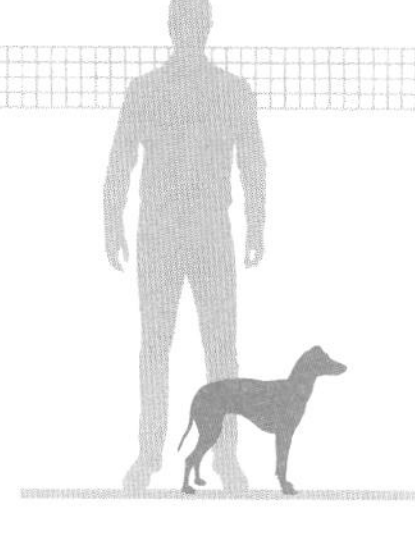

CHILD FRIENDLINESS

GROOMING

FEEDING

EXERCISE

HEIGHT AT SHOULDER
38cm (15 in)

SMOOTH FOX TERRIER

This terrier was bred to go underground and drive foxes out from their lairs. Many breeds, including Bull Terriers, Beagles and Greyhounds (see pp. 88–9 and 232–3), contributed to the Smooth Fox Terrier's development, but the major influence was the Black and Tan Terrier. Initially, in the mid-1800s, the wirehaired and smooth-haired forms were grouped together. By the early 20th century, the Smooth Fox Terrier had become one of Britain's most popular breeds, but since then, its popularity has declined dramatically.

PERSONALITY

The breed's decline can be linked directly to its temperament. These Fox Terriers display a strong independent streak and are not therefore easily trained. They are lively companions, but may sometimes be snappy by nature.

HEALTH AND CARE

Various congenital problems are known, include a lack of teeth. Symptoms of recessive ataxia, another inherited problem, emerge in puppies at between 2–6 months. This causes a degeneration of the spinal cord, and affected dogs lose the ability to walk. Unfortunately, there is no treatment.

AS AN OWNER

The Smooth Fox Terrier is an independent breed that will want to explore on its own when out. Take particular care in areas where there might be underground burrows, because your terrier may go to ground. Fitting your dog with a tracking collar will allow you to locate your pet should it disappear.

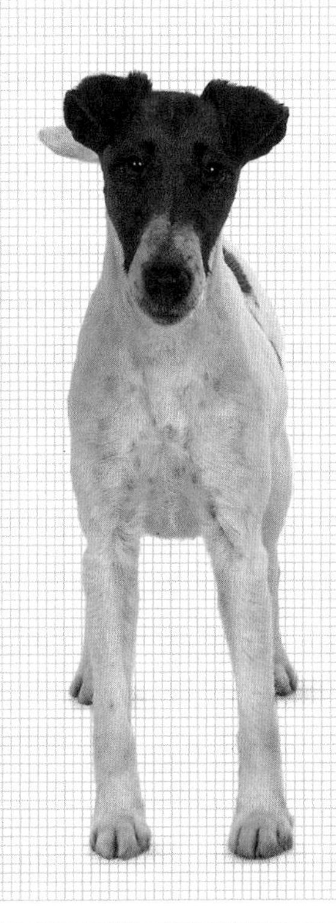

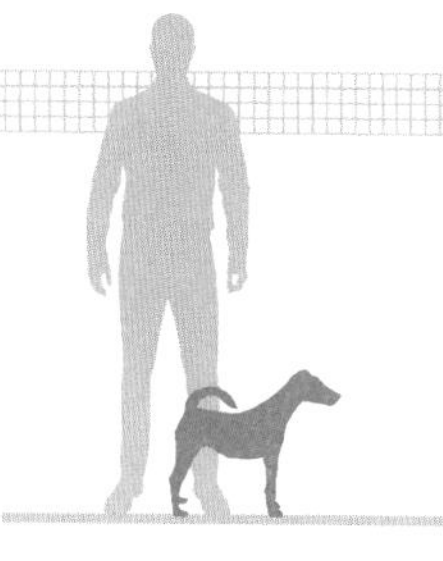

Specification

Predominantly white colouration became a breed characteristic at an early stage, preventing these terriers from being confused with foxes when they emerged from below ground. The coat is smooth and lies flat, but has a relatively hard texture.

RECOGNITION Te
North America, Britain and FCI member countries

LIFESPAN
11–13 years

COLOUR
Mainly white, broken by black or tan

HEAD
Head narrows between the eyes, with the muzzle tapering

EYES
Small, dark and virtually circular

EARS
V-shaped and small; dropping forwards close to the cheeks

CHEST
Deep, but not broad

TAIL
Strong and high-set

WEIGHT
8–9kg (18–20 lb)

CHILD FRIENDLINESS

GROOMING

FEEDING

EXERCISE

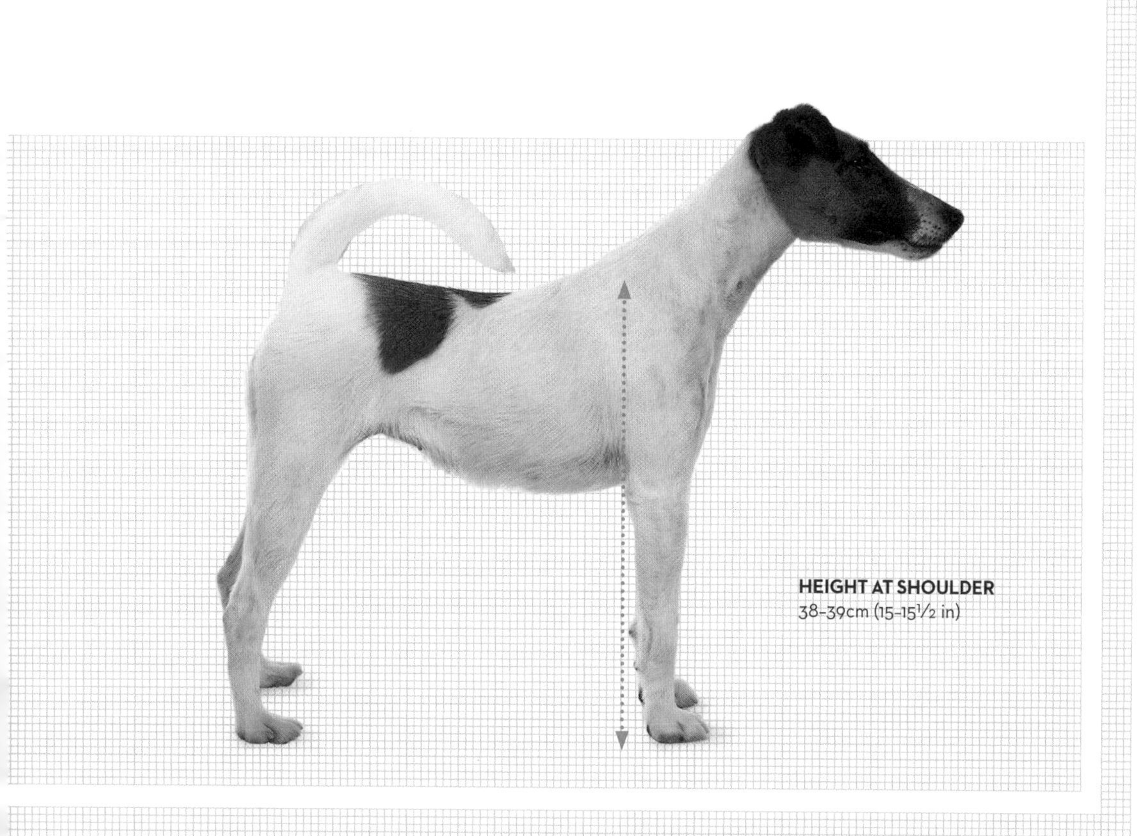

HEIGHT AT SHOULDER
38–39cm (15–15½ in)

WELSH TERRIER

The friendly disposition of these terriers underlies their popularity. Resembling a smaller version of the Airedale (see pp. 164–5), the Welsh Terrier is a much more tolerant by nature. It is probably also the easiest of the terriers to train. This breed is descended principally from the Black and Tan Terrier, having being bred in Wales since the 18th century. Some subsequent crossings with both Lakeland and Airedale Terriers also played a part in its development. Welsh Terriers were expected to go underground and drive out game such as foxes.

PERSONALITY

Lively and bold by nature, the Welsh Terrier is generally a more easy-going household companion than some of the more assertive members of the terrier group. As such, it usually gets along well with other dogs.

HEALTH AND CARE

Welsh Terriers are a hardy, healthy breed. However, these dogs must be immunized regularly against leptospirosis. This potentially deadly infection carried by rats causes liver and kidney failure.

AS AN OWNER

The Welsh Terrier is very playful and is happy to chase after and catch toys. Hand-stripping – or possibly trimming a dog that is not shown – will be necessary about twice a year, and brushing will keep the harsh, wiry coat immaculate.

Specification

Originally called the Welsh Black and Tan Rough-coated Terrier, the black area covers the back, from the neck to the tail, forming what is often described as a jacket. In some cases, however, the black colouring is replaced by grizzled (bluish-grey) colouration.

RECOGNITION Te
North America, Britain
and FCI member countries

LIFESPAN
11–13 years

COLOUR
Black and tan, or grizzle and tan

HEAD
Rectangular, with the muzzle being about half as long as the head

EYES
Set deep, small and dark brown

EARS
Small and V-shaped; fold just above the top of the head

CHEST
Medium-width, with a deep brisket

TAIL
High-set and carried upright

WEIGHT
9kg (20 lb)

CHILD FRIENDLINESS

GROOMING

FEEDING

EXERCISE

HEIGHT AT SHOULDER
Dogs 38–39cm (15–15½ in); bitches sometimes smaller

BOSTON TERRIER

The breed's founder was a dog named Judge, born in the English port of Liverpool. Bred from an English Bulldog and an English White Terrier, he was taken to Boston in the early 1870s. Other breeds that later contributed to the Boston Terrier's development there included the Boxer and the Bull Terrier. Initially, the new breed was used in dog fights, but it entered the show ring in the 1890s. By the 1950s, it had become the most popular breed in North America.

PERSONALITY

In spite of its combative origins, the Boston Terrier today is not aggressive towards other dogs. It is an intelligent, responsive breed and makes an excellent companion.

HEALTH AND CARE

Their large heads can cause puppies to become trapped in the birth canal, so litters may have to be born by Caesarean section. Boston Terriers can also suffer hereditary issues, like cleft palates or lips and various heart defects. Their short coat makes grooming easy.

AS AN OWNER

Try to discourage your pet from plunging into undergrowth, because it can easily injure its thin ears or its prominent eyes here.

Specification

The broad square head is free of wrinkles and is topped with prominent upright ears. The breed is traditionally brindle and white, mirroring that of Judge himself, although a wider range of colours is now accepted. Boston Terriers have a compact appearance with strong limbs and move in a straight, sure-footed way.

RECOGNITION NS
North America, Britain and FCI member countries

LIFESPAN
9–11 years

COLOUR
Black, brindle or seal (black with red highlights), with white markings

HEAD
Square, with a flat top and a well-defined stop; short square muzzle

EYES
Large and round

EARS
Small and erect

CHEST
Wide and deep

TAIL
Short, set low and tapers, not being raised above the horizontal

WEIGHT
Categories extend from under 7kg (15 lb) to over 11.5kg (25 lb)

CHILD FRIENDLINESS

GROOMING

FEEDING

EXERCISE

HEIGHT AT SHOULDER
38–43cm (15–17 in)

STANDARD POODLE

Poodles are bred in three sizes, with the standard being the oldest variety, dating back to the 15th century. Its name comes from the German word 'pudel', which means 'to splash in water'. It was used originally to retrieve shot waterfowl from lakes and rivers, and later became popular as circus performer.

PERSONALITY

Playful and responsive, the Poodle is an excellent companion, still ready to plunge into water if given the opportunity. It is easily trained, active by nature, and will enjoy retrieving games.

HEALTH AND CARE

Standard Poodles are robust dogs, and tend to be fitter overall than the smaller varieties, with few inherited problems. The tear glands, which open on to the inner surface of each lower eyelid, may develop a drainage problem, causing tear-staining at the corners of the eyes. Epilepsy can be a problem in some bloodlines.

AS AN OWNER

Opt for the easily managed sheep-like sporting clip, with the pompon on the tail. The Poodle does not shed its coat, making it ideal for the house-proud!

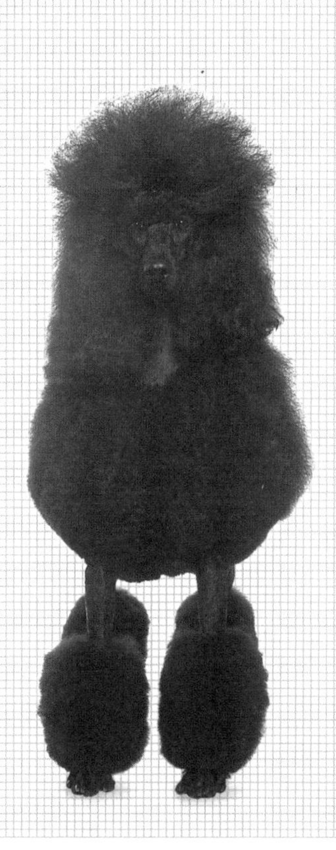

Specification

Its coat is traditionally clipped in one of a range of over 50 variations. The show clip looks decorative but has a practical basis, intended both to keep the dog warm while swimming and to reduce its drag in the water. Hair is left to form 'bracelets' on the lower limbs, to protect against rheumatism, and the pompon on the tail is used to locate the dog in water.

RECOGNITION NS
North America, Britain and FCI member countries

LIFESPAN
11–13 years

COLOUR
Solid colours, from white, cream and apricot through to blue and black

HEAD
Long and fine, with a relatively narrow skull

EYES
Almond-shaped and dark

EARS
Long, wide and set low, close to the sides of the face

CHEST
Deep and wide

TAIL
Set high, thick at its base, and carried away from the body

WEIGHT
20.5–32kg (45–70 lb)

HEIGHT AT SHOULDER
38cm (15 in) or more

ENGLISH COCKER SPANIEL

English Cocker Spaniels are popular in the show ring and as household pets, plus they are also adept in field trials. The Cocker and Springer Spaniels (see pp. 128–9 and 146–7) share a common ancestry. Both types were originally described as Land Spaniels, but gradually a divergence in type emerged, with the English Cocker Spaniel being recognizable by its smaller size.

PERSONALITY

Friendly and responsive, the English Cocker Spaniel is a lively breed with an enthusiastic outlook on life. Thanks to their intelligence, they can be trained quite easily, being keen to please and responding well to encouragement.

HEALTH AND CARE

Although usually a genial dog, there is a disturbing inherited condition known as the Rage Syndrome that occasionally afflicts solid (single)-coloured English Cocker Spaniels. Such dogs suddenly and inexplicably become aggressive, then revert almost immediately to their previously docile personalities.

AS AN OWNER

These spaniels are full of energy, making them a good choice as a family pet alongside older children, but plenty of grooming is also required. They need plenty of attention and opportunities to play to prevent them from becoming bored and destructive.

Specification

The English Cocker Spaniel creates an impression of greater power and strength than the now-distinct American breed, being both taller and heavier. It is not especially fast, but appears to move without obvious effort, propelled along by its powerful hind legs. The tail is usually kept horizontal, but may occasionally be lifted slightly if the dog is especially excited.

RECOGNITION S
North America, Britain and FCI member countries

LIFESPAN
10–12 years

COLOUR
Solid red, liver or black, and parti-colours, including roans, are permitted, as are tan markings

HEAD
Arched and flattened skull

EYES
Well-spaced, with a slight oval shape

EARS
Long and set low, lying close to the sides of the face

CHEST
Deep

TAIL
Carried horizontally and moving repeatedly when the dog is working

WEIGHT
Dogs 12.5–15.5kg (28–34 lb); bitches 11.75–14.5kg (26–32 lb)

CHILD FRIENDLINESS

GROOMING

FEEDING

EXERCISE

HEIGHT AT SHOULDER
Dogs 40.5–43cm (16–17 in); bitches 38–40.5cm (15–16 in)

BASENJI

These unusual dogs originate from Central Africa, having been kept as hunting companions for centuries. They are highly responsive, intelligent and loyal, and possess various unusual mannerisms – including catlike behaviour. They spend a lot of time grooming themselves, are agile climbers, and may also spend long periods simply watching from a window. A pair of Basenjis was brought to England for the first time in 1895, but died from distemper. The breed was then reintroduced to Britain in 1937, reaching America soon afterwards, but sadly, all but one of these Basenjis again succumbed to this disease. Gradually, however, Western bloodlines have evolved.

PERSONALITY

Always alert, the Basenji bonds well with people. It has a friendly nature towards family members, but is often reserved with strangers. Although it does not bark, it is quite capable of growling, and has an unusual yodelling call when excited or playing.

HEALTH AND CARE

The first Basenjis kept outside Africa proved to be very delicate, encountering diseases to which they had no natural immunity. Even today, it is particularly important to keep their immunizations current. Puppies should be checked for umbilical hernias while older dogs may be susceptible to digestive upsets, which afflict certain bloodlines.

AS AN OWNER

Basenjis only have one period of heat annually, rather than two, like most breeds.

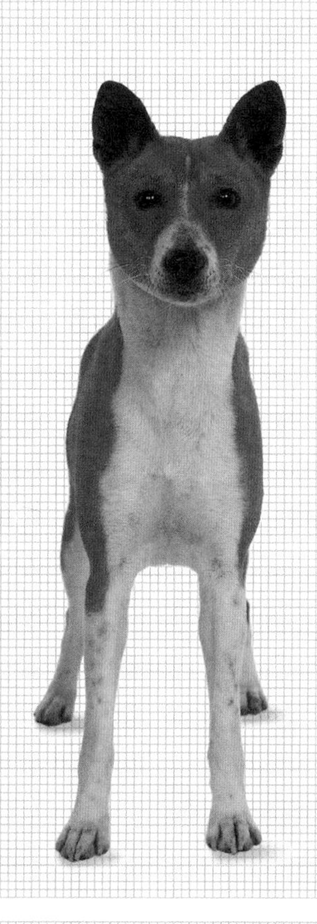

Specification

The wrinkled forehead – especially evident in puppies – gives the Basenji a worried look. Originating close to the Equator, its coat is short and fine. These hounds have white markings on the chest, the underparts of the body, at the tip of the tail and often a white blaze between the eyes.

RECOGNITION H
North America, Britain and FCI member countries

LIFESPAN
11–13 years

COLOUR
Black or red with white markings; plus brindles and tricolours

HEAD
Relatively short muzzle

EYES
Dark-brown to hazel

EARS
Small, erect and set forwards on top of the skull

CHEST
Medium-width

TAIL
Bends forwards, curled to one side

WEIGHT
Dogs 10kg (22 lb); bitches 9kg (20 lb)

CHILD FRIENDLINESS

GROOMING

FEEDING

EXERCISE

HEIGHT AT SHOULDER
Dogs 43cm (17 in); bitches 40.5cm (16 in)

KEESHOND

This breed has become recognized as the national breed of Holland, having traditionally lived on canal barges, guarding their cargo, and it is sometimes still known as the Dutch Barge Dog. Its name is pronounced as 'kays-hawnd', and the plural form is Keeshonden. The Keeshond probably gained its name from the nickname – Kees – of Cornelis de Gyselaer, the leader of an uprising in Holland against the established Royal House of Orange. De Gyselaer used his dog as his party's symbol and the breed fell out of favour for a period after the revolt was crushed.

PERSONALITY

Intelligent and watchful, the Keeshond makes a loyal companion, although it can be noisy and energetic. It is very friendly as a family pet and settles down happily in the home.

HEALTH AND CARE

Heart defects, especially those affecting the mitral valve, can be a congenital weakness. The breed can also suffer from inherited epilepsy, which may not develop until the dog is over 3 years old. Ongoing medication is usually required to treat it.

AS AN OWNER

Regular grooming is essential, particularly in spring when the thicker winter coat is shed. If you like the ruff, choose a male puppy, as they develop a more evident mane. The characteristic shaded markings become more prominent with age too.

Specification

The Keeshond is a typical spitz breed, although of unknown ancestry, with distinctive wolflike colouring. The narrow muzzle is covered with relatively short hair, and the ears are pricked, highlighting its alert nature. The coat stands up, away from the body, and there is a paler ruff of long fur around the neck.

RECOGNITION NS
North America, Britain and FCI member countries

LIFESPAN
10–12 years

COLOUR
May vary from light to dark, most commonly black and silver

HEAD
Well-proportioned with a medium-length muzzle

EYES
Almond-shaped and dark brown

EARS
Erect, set high and triangular

CHEST
Strong deep chest

TAIL
Curled up and over the back

WEIGHT
25–30kg (55–66 lb)

GROOMING

FEEDING

EXERCISE

HEIGHT AT SHOULDER
Dogs 45.5cm (18 in); bitches 43cm (17 in)

NORWEGIAN BUHUND

The Buhund has been bred in Norway for over 1,000 years, but only since the 1970s has it become popular internationally. Descended from the ancient Iceland Dog, the Buhund was traditionally kept mainly for herding sheep. Its unusual name derives from the addition of the Norwegian word 'bu' - the name given to the temporary shelters built by shepherds in the summer - to 'hund', which means 'dog'. During the winter, the breed also acted as a sled dog and as a companion on hunting expeditions.

PERSONALITY

The Norwegian Buhund has an intelligent, adaptable nature and is a good guard dog, alert to the presence of strangers. Energetic and lively, this breed needs plenty of exercise.

HEALTH AND CARE

Its grooming requirements are not very time-consuming but will take longer in spring, when the dense winter coat is being shed. The breed can suffer from hip dysplasia, so ensure that any puppy you are considering has been bred from stock that has been properly screened for this condition.

AS AN OWNER

Norwegian Buhunds can be trained easily but have a tendency to develop into one-person dogs so, if possible, involve everyone in the family in your pet's care. The breed's working instincts are strong, so as always, keep your pet on a leash around sheep and other farm livestock.

Specification

The Norwegian Buhund still closely resembles its Icelandic ancestor. It has the raised ears typical of a spitz breed, and is protected against the elements by a dense undercoat for insulation, combined with a short, rough-textured top coat.

RECOGNITION He
North America, Britain
and FCI member countries

LIFESPAN
11-13 years

COLOUR
Cream and wheaten shades
through red to black

HEAD
Broad and wedge-shaped, with the muzzle tapering slightly

EYES
Dark and almond-shaped

EARS
Large and rounded at the tips; set well-back

CHEST
Deep and powerful

TAIL
Set high and bushy; curled forward over the back and down the side

WEIGHT
24-26kg (53-58 lb)

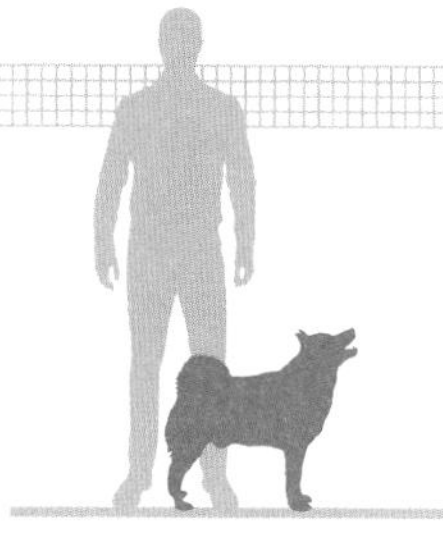

CHILD FRIENDLINESS

GROOMING

FEEDING

EXERCISE

HEIGHT AT SHOULDER
43-45.5cm (17-18 in)

SOFT COATED WHEATEN TERRIER

Originally called the Irish Wheaten Terrier, this breed has proved to be very versatile. As well as hunting game and killing rodents, it can herd livestock and proves an alert guard dog. It is an ideal choice if you like the liveliness of terriers, but are looking for a more tolerant character, especially in terms of mixing well with other dogs. The Soft Coated Wheaten is the oldest of today's four existing Irish terrier breeds, with an ancestry extending back over 200 years.

PERSONALITY

Friendly and self-confident, this breed is hardy, easy to care for and makes a superb companion – it deserves to be more popular as a pet. It will adapt well to various surroundings, although it tends to be happier in a rural environment.

HEALTH AND CARE

The Soft Coated Wheaten Terrier is generally a healthy breed. Although the coat is single-layered, it affords good protection against the elements, with grooming being relatively straightforward. It will take up to 18 months for the coat of Soft Coated Wheaten Terriers to lighten and lose any darker markings, and for its distinctive texture to develop fully.

AS AN OWNER

Although this breed is significantly larger than many other terriers, it can be trained easily. It is playful by nature, and is patient with children. Few breeds are as versatile or better suited to living in a family environment.

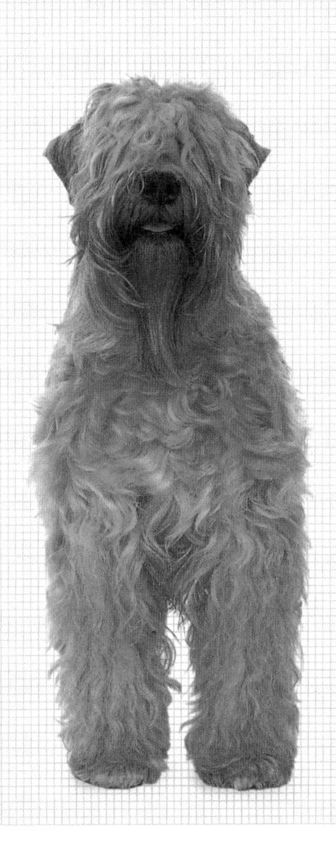

Specification

This medium-sized breed has a soft, silky coat, which is wavy in adults but lies flat in young terriers. The nose is black as are the lips.

RECOGNITION Te
North America, Britain and FCI member countries

LIFESPAN
12–14 years

COLOUR
Clear wheaten

HEAD
Rectangular and long, with a strong muzzle

EYES
Medium-sized and brown

EARS
Small or medium-sized and level with the head

CHEST
Deep

TAIL
Well-set, held upward, not extended over the back

WEIGHT
Dogs 16–18kg (35–40 lb); bitches 13.5–16kg (30–35 lb)

GROOMING

FEEDING

EXERCISE

HEIGHT AT SHOULDER
Dogs 45.5–48cm (18–19 in); bitches 43–45.5cm (17–18 in)

AUSTRALIAN CATTLE DOG

This tough working breed is still kept for herding cattle on ranches in Australia, but has now built up an international reputation in the show ring. Its extremely high energy levels make it unsuitable for living in a city unless you can provide it with plenty of regular exercise off the leash. Herding dogs brought from Britain and elsewhere did not adapt well to the harsh climate of Australia. Stockmen tried crossing various breeds with the native dingo, ultimately giving rise to the Australian Cattle Dog. Its distinctive colouring can be traced back to two blue merle (a marbled effect created when separate hair colours mix together) Smooth Collies (see pp. 184–5) brought from Scotland in 1840. Puppies may still be born white at birth, reflecting an early Dalmatian input to the breed.

PERSONALITY

These dogs are active and can easily become bored. They form a strong bond with their owners.

HEALTH AND CARE

Hardy and long-lived, it is a member of this breed that is the oldest dog on record. Bluey survived for over 29 years and worked for 16 of them. Grooming needs are minimal.

AS AN OWNER

Careful socialization, both with people and other dogs is required, and these dogs must be trained to curb their powerful herding instincts.

Specification

Stocky, with a powerful head and upright ears, the Australian Cattle Dog often has a blue mottled coat. The colouring of individual dogs varies widely, allowing them to be distinguished from one another easily at a distance.

RECOGNITION He
North America, Britain and FCI member countries

LIFESPAN
14–20 years

COLOUR
Blue, sometimes with mottling and speckled red

HEAD
Broad, with a definite stop; powerful, medium-length muzzle

EYES
Medium-sized, oval and dark brown

EARS
Relatively small, widely spaced and pricked

CHEST
Muscular and deep

TAIL
Set moderately low, curving slightly at rest; must not extend vertically above its base

WEIGHT
16–20.5kg (35–45 lb)

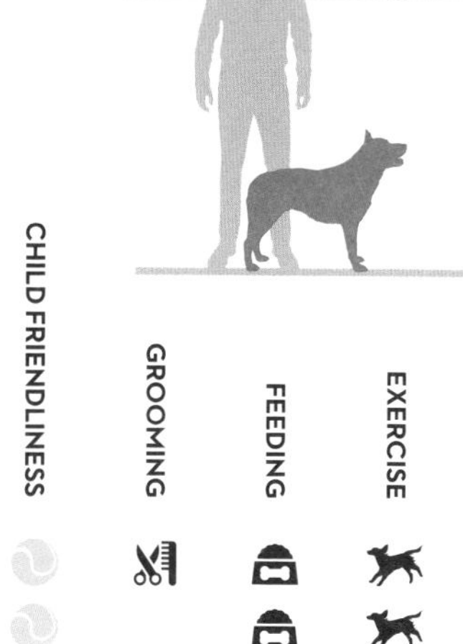

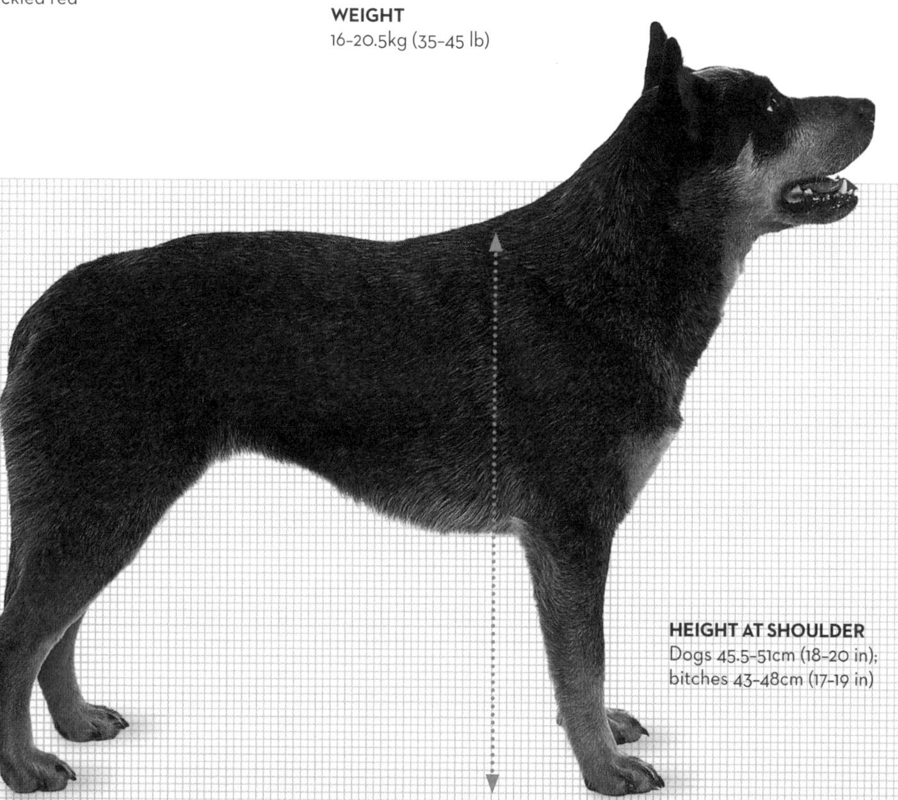

HEIGHT AT SHOULDER
Dogs 45.5–51cm (18–20 in); bitches 43–48cm (17–19 in)

WELSH SPRINGER SPANIEL

This breed has been popular in Wales for over a millennium and still remains numerous there today. It may be the ancestral form of all the British spaniels. Its origins are unclear, although they may be linked with those of the Brittany (see pp. 132–3) from northern France. Until the late 1800s, the breed was often called the Welsh Cocker Spaniel, although it is significantly heavier than its English counterpart. Welsh Springers were first introduced to North America in the late 1800s.

PERSONALITY

The Welsh Springer is a breed that is eager to please. It is lively, active and affectionate, thriving in rural surroundings where it can locate and flush game in its traditional working style.

HEALTH AND CARE

The coat provides excellent protection, but always groom your pet after country walks or working excursions, to remove any burrs and thorns, checking for ticks too.

AS AN OWNER

The Welsh Springer was developed to work outdoors and is unsuited to urban living. Good training is essential to prevent the dog following its nose and disappearing after a scent when out for a walk, but these spaniels learn readily. They will settle well in a home, provided they can be given enough exercise, and also make good guard dogs.

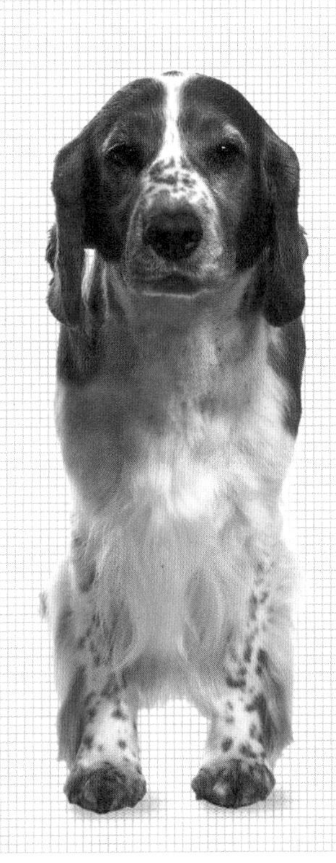

Specification

The dark red and white colouring of the Welsh Springer Spaniel is distinctive, with flecking – the intermingling of red and white hairs – sometimes occurring too. The legs and underparts are heavily feathered and there is lighter feathering on the ears and tail. The Welsh Springer has a narrower head than most spaniels.

RECOGNITION S
North America, Britain and FCI member countries

LIFESPAN
11–13 years

COLOUR
Red and white

HEAD
Domed and medium-length

EYES
Dark and oval

EARS
Lie close to the cheeks, set at the level of the eyes

CHEST
Muscular, extending level with the elbows

TAIL
Extends the topline, lying virtually horizontal and raised slightly when excited

WEIGHT
16–20.5kg (35–45 lb)

CHILD FRIENDLINESS

GROOMING

FEEDING

EXERCISE

HEIGHT AT SHOULDER
Dog 45.5–48cm (18–19 in); bitches 43–45.5cm (17–18 in)

SCHNAUZER

The Schnauzer makes an ideal family pet – being lively and social with family members and thriving in the company of children, with their size also being in their favour. Schnauzers have a long ancestry dating back more than 600 years, having been kept originally to hunt rats. They used to be called Wire-haired Pinschers; but Schnauzer, meaning 'whiskered snout', was adopted in 1879.

PERSONALITY

Bold and dependable, the Schnauzer is a quick learner, thanks to its natural intelligence. It forms a strong bond with people in its immediate circle, and a Schnauzer puppy will soon settle well as a member of the family.

HEALTH AND CARE

Puppies may occasionally lack drainage holes in their tear glands, meaning that tear fluid cannot drain normally out of the eyes and so overflows down the face. The problem can be corrected with surgery.

AS AN OWNER

Coat care requires professional grooming because Schnauzers do not shed. Be prepared for a lively dog that wants to be constantly involved in family life, participating in whatever is happening and enthusiastic about games.

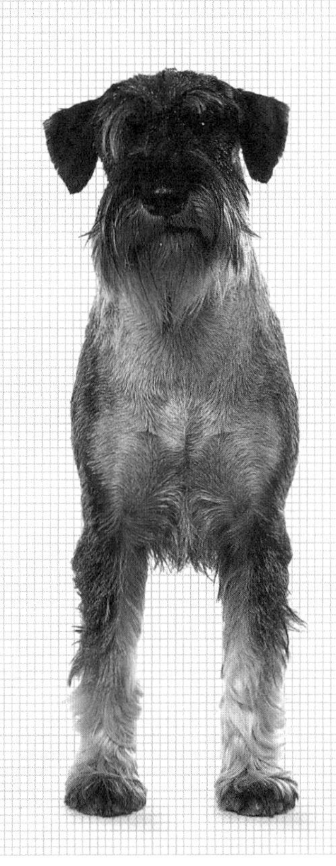

Specification

Rough-coated, some of these Schnauzers are an unusual mixed grey colour known as 'salt and pepper', which can vary significantly in depth between individuals. The coat should be dense, with a tight, wiry texture. The outer coat is raised away from the body, measuring up to 5cm (2 in) long on the back.

RECOGNITION W
North America, Britain and FCI member countries

LIFESPAN
11–13 years

COLOUR
Salt and pepper (greys ranging from a silvery shade to iron grey) or solid black

HEAD
Rectangular and powerful, with the muzzle matching in length and width

EYES
Oval, dark brown and medium-sized

EARS
V-shaped, set high; and carried so the inner edges lie close to the cheeks

CHEST
Medium-width

TAIL
Set high and carried erect

WEIGHT
Dogs 16–20.5kg (35–45 lb); bitches 13.5–18kg (30–40 lb)

CHILD FRIENDLINESS

GROOMING

FEEDING

EXERCISE

HEIGHT AT SHOULDER
Dogs 47–49.5cm (18½–19½ in); bitches 44.5–47cm (17½–18½ in)

BRITTANY

This breed arose in northern France, in western Brittany. It was first kept as a spaniel, used to flush and retrieve game. It then evolved into a true all-purpose gundog, following crossbreeding with setters and pointers. This is why the qualifying description of 'spaniel' has been dropped from its name. The Brittany was first seen in North America in 1931.

PERSONALITY

Responsive and adaptable, the Brittany makes an affectionate companion. It has a determined side to its nature, combined with considerable stamina, but is usually keen to please, and it can be trained easily.

HEALTH AND CARE

Some bloodlines are afflicted by the blood-clotting disorder known as haemophilia. If you are acquiring a Brittany, make sure the breeding stock has been screened. Care of the Brittany is straightforward and its coat is easy to groom, compared with those of true spaniels.

AS AN OWNER

Brittanys need regular, off-leash exercise in the countryside. Always inspect your dog if it has been in undergrowth, because it may pick up burrs or ticks, and check the feet for embedded grass seeds, which could puncture the skin and cause infection.

Specification

The colour of the Brittany is now more restricted in North America than in Britain, where the breed was only recognized in 1976. It has a fine, dense coat, with little feathering on the backs of the legs or fringing on the ears. Another point of distinction with true spaniels is that the lips are tighter, not tending to droop down.

RECOGNITION S
North America, Britain and FCI member countries

LIFESPAN
11–13 years

COLOUR
Orange or liver and white, and maybe roan; any black markings are discouraged

HEAD
Slightly rounded and medium-length skull

EYES
Set deep and emphasized by the eyebrows

EARS
Triangular and short with rounded tips, and lie flat

CHEST
Deep, extending down to the elbows

TAIL
Set high, often naturally tail-less or extremely short

WEIGHT
13.5–18kg (30–40 lb)

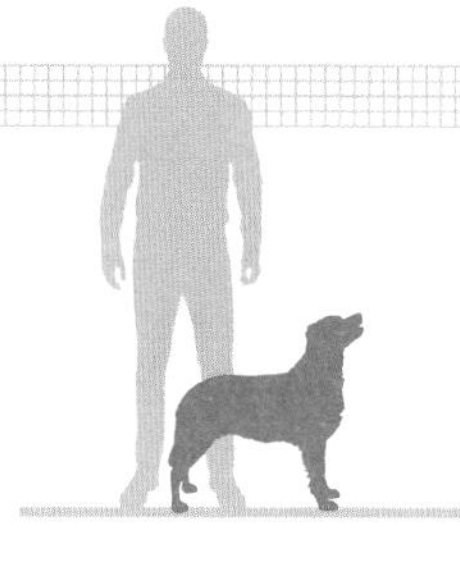

CHILD FRIENDLINESS

GROOMING

FEEDING

EXERCISE

HEIGHT AT SHOULDER
44.5–52cm (17½–20½ in)

KERRY BLUE TERRIER

Originating in County Kerry, this breed is officially regarded as Ireland's national dog. The Kerry Blue is a versatile terrier, and has also worked effectively as both a retriever and a herder, in addition to hunting vermin. Its origins are unknown, but it may be descended from indigenous Irish terrier breeds, perhaps crossed with Bedlington Terriers. It became popular in iIreland during the late 1800s, and had reached America by the 1920, but has since faded in popularity.

PERSONALITY

An intelligent and adaptable breed with strong working instincts and a playful side too.,the Kerry Blue revels in human company and is keen to learn.

HEALTH AND CARE

Puppies typically aged 9–16 weeks occasionally develop signs of a nervous system disorder, resulting in tremors of the head and a stiffness of the forelegs, leading to paralysis. Tumours of the hair follicles may also occur in older individuals. These appear as swellings, which become noticeable during grooming.

AS AN OWNER

Be careful with a Kerry Blue Terrier around cats, as this breed often displays a strong dislike of them that is hard to train out. Kerries are patient with children, however, and make good watchdogs.

Specification

Its coat colour varies from blue-grey to grey-blue. This should be even, but the extremities – muzzle, ears, and feet – are often darker, or even black. The coat must be dense, soft, and wavy. Puppies are black at birth, but their colour lightens over the first year or so: a process described as 'clearing'.

RECOGNITION Te
North America, Britain and FCI member countries

LIFESPAN
13-15 years

COLOUR
Blue-grey

HEAD
Long with the muzzle matching the skull length

EYES
Small and dark

EARS
Small, V-shaped and lying close to the cheeks

CHEST
Deep and broad

TAIL
Medium-length, straight and carried above the back

WEIGHT
15-18kg (33-40 lb), with bitches being at the lighter end of the spectrum

CHILD FRIENDLINESS

GROOMING

FEEDING

EXERCISE

HEIGHT AT SHOULDER
Dogs 45.5-49.5cm (18-19½ in); bitches 44.5-48cm (17½-19 in)

SHAR-PEI

The survival of the ancient Chinese Shar-Pei and its widespread international popularity today are thanks to a Hong Kong dog fancier. In the 1970s he wrote an article about the plight of this ancient Asian breed, when it had almost become extinct. As a result, some American enthusiasts imported stock and breeding began to save the Shar-pei. The broad square head is free of wrinkles and is topped with prominent upright ears located at the corners of the skull.

PERSONALITY

A bold, fearless breed, the Shar-pei can be aggressive towards other dogs.

HEALTH AND CARE

At one point the Shar-pei was regarded as the rarest breed of dog in the world, reduced to a total of fewer than 60 individuals, so today's stock is built on a small gene pool. Excessive folds of skin may trigger localized infections, particularly during hot weather. Problems affecting the eyes, again because of the folds of skin around them, can also arise.

AS AN OWNER

Coat length varies markedly. Shorthaired Shar-peis have so-called 'horse coats', whereas those about 2.5cm (1 in) in length are described as 'brush coats'. This breed can easily develop a dominant nature without proper training.

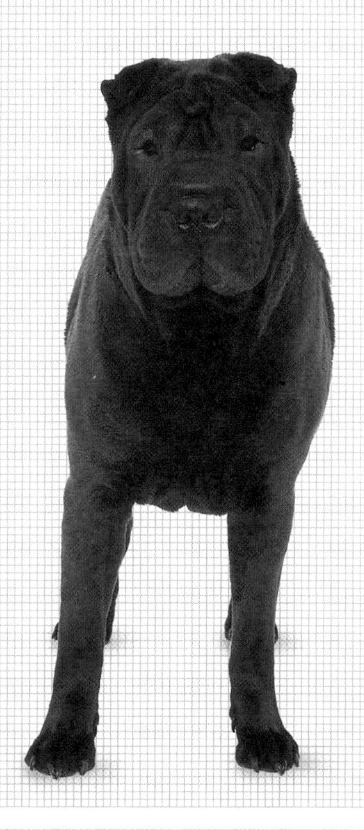

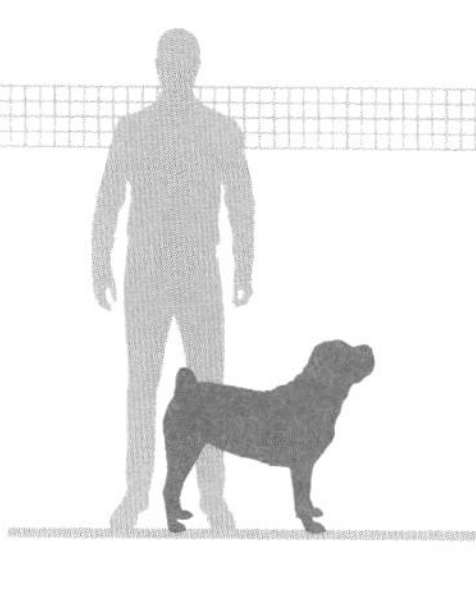

Specification

Puppies have deep, loose skin folds on their bodies, which become less conspicuous as the dog literally grows into its skin The breed's name means 'sandpaper-like coat', describing its rough, bristly texture. Another highly unusual feature of the Shar-Pei is its blue tongue.

RECOGNITION NS
North America, Britain and FCI member countries

LIFESPAN
10–12 years

COLOUR
Any solid colour apart from albino

HEAD
Large, with a distinctively broad muzzle

EYES
Small, sunken and almond-shaped

EARS
Set high, widely spaced and usually lie flat

CHEST
Deep and broad

TAIL
Set high on the back, round at the base and tapers to a point; curled over or to the side of the back

WEIGHT
20.5–27kg (45–60 lb)

CHILD FRIENDLINESS

GROOMING

FEEDING

EXERCISE

HEIGHT AT SHOULDER
45.5–51cm (18–20 in)

WHIPPET

Often described as the poor man's racehorse, the Whippet can sprint at speeds of up to 56kph (35 mph) for short distances. Today, their responsiveness makes them an ideal choice for obedience competitions, while their natural athleticism means the breed often excels in more energetic activities, including fly ball and agility contests. Whippets were created in the north of England, with the Greyhound (see pp. 232–3) being involved in their ancestry. Terriers – particularly Bedlingtons, which also originated in this part of the world – and some form of spaniel also probably had an input.

PERSONALITY

Potentially shy and sensitive, Whippets are nevertheless very affectionate towards people they know well. They love playing games, and they make good retrievers when chasing after balls.

HEALTH AND CARE

Whippets are low-maintenance dogs, with coats that need very little grooming, but they do feel the cold. Males can occasionally suffer from cryptorchidism, when one or both testes fail to descend normally into the scrotum.

AS AN OWNER

A Whippet will need space to run, ideally in open countryside, rather than just in an urban park. They may dig in the garden at home, and fences must be both high and secure, because Whippets can jump well.

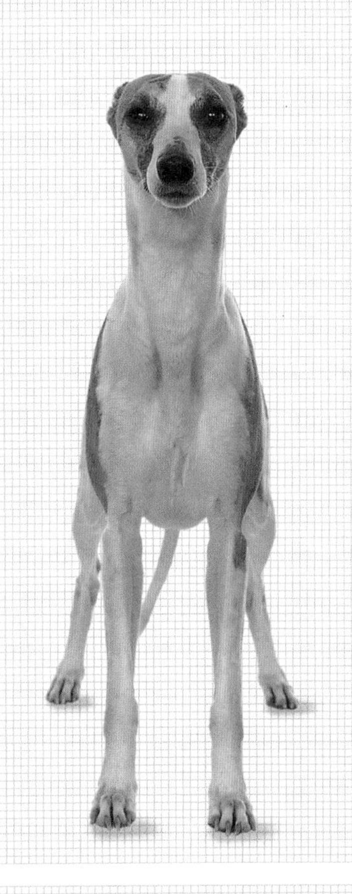

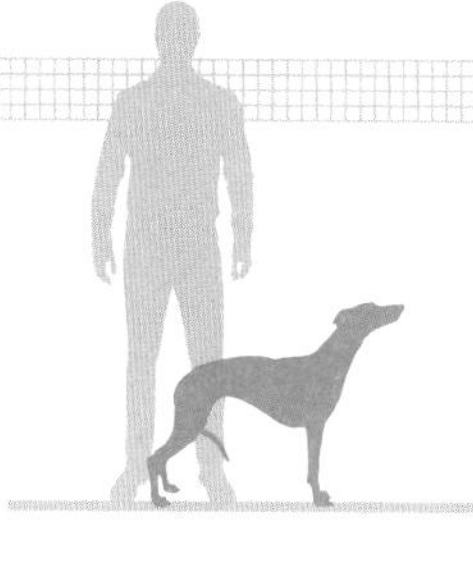

Specification

A whippet has a slender, deep-chested body and a so-called 'roach back' that curves down to the tail. Their coats are sleek, smooth and short, reinforcing the well-muscled appearance of the breed, which is especially apparent over the hindquarters.

RECOGNITION H
North America, Britain and FCI member countries

LIFESPAN
13–15 years

COLOUR
Any colour

HEAD
Long and lean, with a black nose

EYES
Must be large, dark and of the same colour.

EARS
Small, rose ears, which are folded back along the neck when dog is resting

CHEST
Deep, reaching almost down to the elbow, with well-sprung ribs behind

TAIL
Not carried higher than top of the back

WEIGHT
Approximately 12.5kg (28 lb)

CHILD FRIENDLINESS

GROOMING

FEEDING

EXERCISE

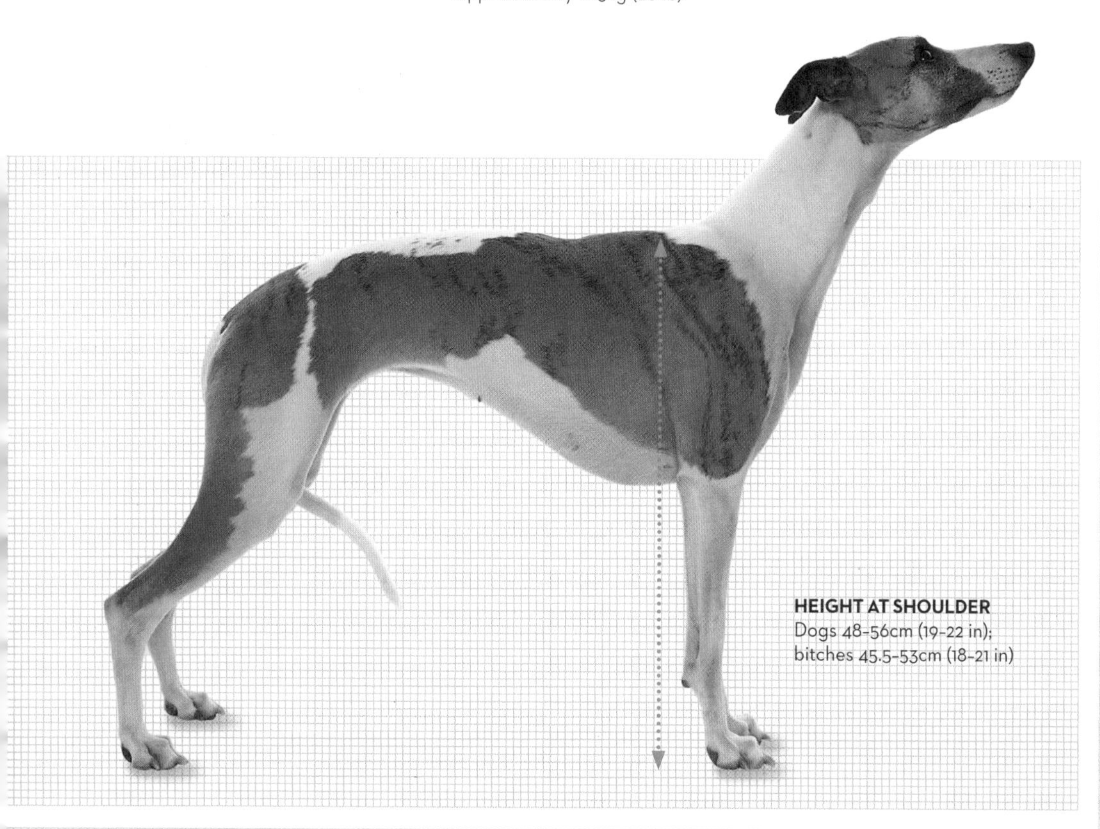

HEIGHT AT SHOULDER
Dogs 48-56cm (19-22 in); bitches 45.5-53cm (18-21 in)

BORDER COLLIE

Sheepdogs have existed in Britain since Roman times. The Border Collie was named after the area between Scotland and England where it was first bred. This breed reached North America in the 1880s. Considered to be the most talented and intelligent breed of its type, it excels in sheepdog trials, agility and fly ball competitions. However, if choosing a Border Collie as a pet appeals, then it is essential that you have enough time to dedicate to this demanding breed.

PERSONALITY

Capable of developing an intuitive relationship with its owner when working, this breed is highly responsive and intelligent, learning quickly and possessing great reserves of stamina. It is also affectionate towards members of its family circle.

HEALTH AND CARE

These collies are susceptible to a hereditary eye condition called progressive retinal atrophy (PRA), which ultimately causes blindness. Symptoms only emerge when dogs are 3–5 years old. Breeding stock should be checked accordingly.

AS AN OWNER

Border Collies require adequate physical and mental exercise each day, and can become neurotic. These dogs have strong working instincts, and should be kept on a leash around sheep especially.

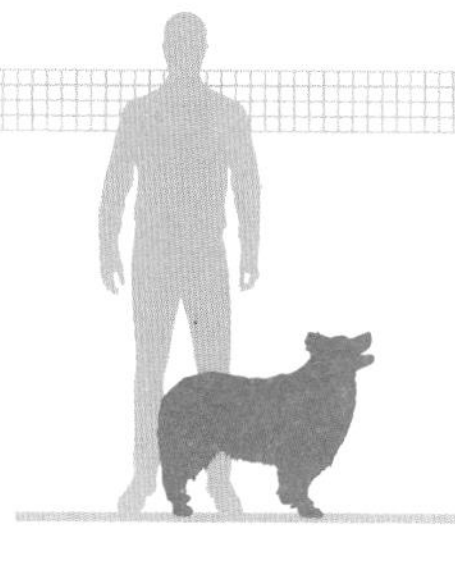

Specification

In the past, its working ability was considered to be the only relevant attribute, but since being recognized for show purposes – it was acknowledged by the British Kennel Club in 1976 – judging standards have been drawn up for the breed. Both rough- and smooth-coated forms are acceptable, with their gait still considered more significant than their colouration and markings.

RECOGNITION He
North America, Britain and FCI member countries

LIFESPAN
11-13 years

COLOUR
Often black and white; can be tricoloured

HEAD
Flat and medium-width; the muzzle length matches that of the skull

EYES
Oval, medium-sized and well-spaced

EARS
Medium-sized; sometimes held erect or semi-erect

CHEST
Deep, relatively broad, reaching as far as the elbow

TAIL
Set low; carried low when concentrating

WEIGHT
13.5-20kg (30-44 lb)

CHILD FRIENDLINESS

GROOMING

FEEDING

EXERCISE

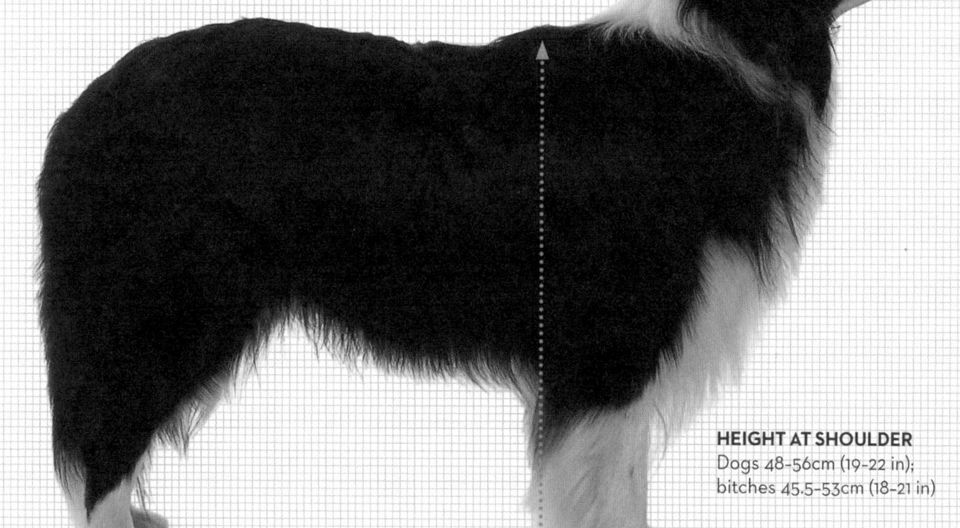

HEIGHT AT SHOULDER
Dogs 48-56cm (19-22 in); bitches 45.5-53cm (18-21 in)

AUSTRALIAN SHEPHERD

This breed's roots lie in the Basque region of the Pyrenees, between France and Spain. Some shepherds from here emigrated to Australia, taking their dogs with them. Subsequently, a second wave of emigration occurred in the late 1800s, this time from Australia to North America, and again these dogs moved with their owners, with the breed evolving further, particularly in California. The Australian Shepherd has a keen natural intelligence and acute powers of observation. Today these versatile dogs still work on American ranches, but they are also employed with disabled people, for search-and-rescue work, and as sniffer dogs to detect illegal drugs.

PERSONALITY

Friendly, tolerant and quiet, the Australian Shepherd has responsive, dependable nature, plus plenty of energy for play. It is usually good with children.

HEALTH AND CARE

Merle to merle breedings should never be permitted, because of sight and/or hearing loss.

AS AN OWNER

This breed has a double-layered, weather-resistant coat. Its grooming needs increase during the spring, when much of the undercoat is shed.

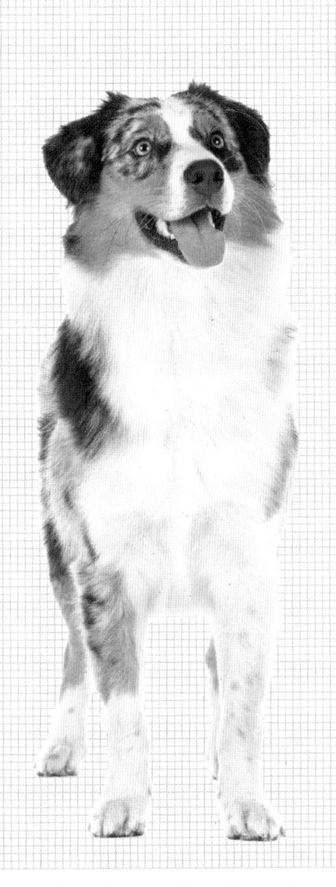

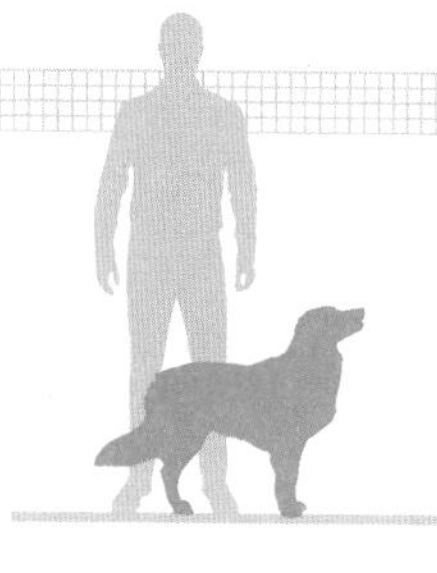

Specification

The Australian Shepherd has the typical attributes of most sheepdogs, with a slightly tapering muzzle and ears set high on the head, creating an alert impression. The eyes can be blue as well as brown, and the coat may be solid red, liver or black, plus blue or red merle, both of which darken with age. White and tan markings are also allowed.

RECOGNITION He
North America, Britain and FCI member countries

LIFESPAN
11–13 years

COLOUR
Black, red, blue or red merle, with or without white markings and/or tan points

HEAD
Flattish, with a medium-length muzzle

EYES
almond-shaped; brown, amber, blue, or any combination of these colours

EARS
Set high; triangular and medium-sized

CHEST
Deep but not broad, reaching to the elbows

TAIL
Straight, sometimes naturally bobbed

WEIGHT
16–32kg (35–70 lb)

CHILD FRIENDLINESS

GROOMING

FEEDING

EXERCISE

HEIGHT AT SHOULDER
Dogs 51-58.5cm (20–23 in); bitches 45.5-53cm (18–21 in)

CHOW CHOW

The unusual name of this breed may come either from a Chinese description for ship's cargo or the Cantonese word 'chow', meaning 'edible'. Descended from the Han Dog, which was known over 2,000 years ago, the Chow Chow was kept primarily, but not exclusively, to be eaten. Young puppies were carefully reared as food on a diet of grain. Chow Chows were also used to pull carts and to act as guard dogs. One Emperor of China even kept 5,000 Chow Chows to serve as hunting companions. The breed reached the West in 1780.

PERSONALITY

The breed is highly intelligent and is not instinctively playful. It does not take to strangers readily either.

HEALTH AND CARE

Some Chow Chows are born with tails that are significantly shorter than normal, although they are otherwise healthy. The breed may also have problems affecting the eyelids, including entropion and additional rows of eyelashes, which are likely to require surgical correction.

AS AN OWNER

The Chow Chow is one of the hardest breeds to train successfully, possessing a strong independent streak. The smooth-coated variety requires less grooming. Chow Chows have deep-set eyes that limit their peripheral vision, so may prove nervous if approached from the side.

Specification

The ruff of longer fur around the head has been likened to a lion's mane. There are two recognizable coat types and the mane is seen only in the rough-coated form. The Chow Chow is a stoutly built, muscular breed, with relatively small ears and a bushy tail that curls over the back.

RECOGNITION NS
North America, Britain and FCI member countries

LIFESPAN
11-13 years

COLOUR
Cream, red, cinnamon, blue or black

HEAD
Broad skull and broad, short muzzle

EYES
Deep-set, widely spaced; dark brown

EARS
Triangular, with slightly rounded tips; erect, but tilt slightly forwards

CHEST
Deep, muscular and broad

TAIL
Set high on the back and carried low over the back

WEIGHT
22.5-32kg (50-70 lb)

CHILD FRIENDLINESS

GROOMING

FEEDING

EXERCISE

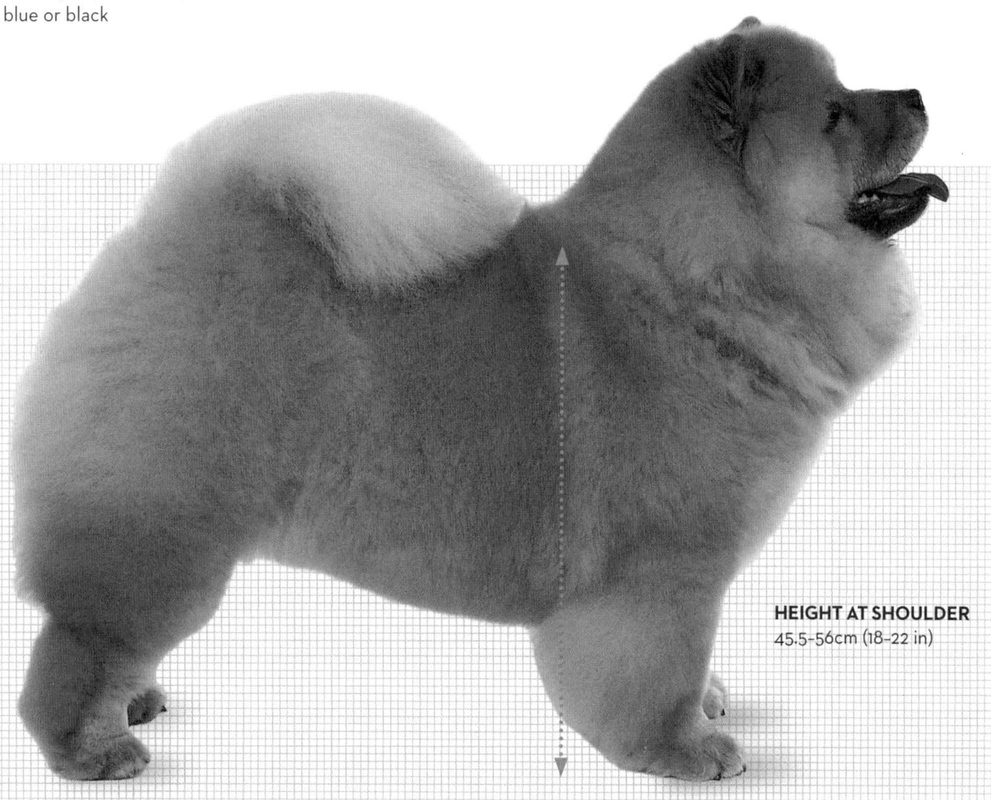

HEIGHT AT SHOULDER
45.5-56cm (18-22 in)

ENGLISH SPRINGER SPANIEL

The description 'springer' relates not to the jumping abilities of these spaniels, but to the way in which they work, 'springing' game out of the undergrowth. This breed was known originally as the Wood Spaniel and possesses tremendous energy and stamina. It has existed in a recognizable form for over 350 years, becoming well known in North America during the 1920s.

PERSONALITY

Exuberant by nature, this breed is an industrious worker and will always be keen to please. It relates well to people and makes a lively companion. Its powers of concentration, whether learning or working, are good.

HEALTH AND CARE

This breed is particularly susceptible to a range of congenital eye problems. It can suffer from inward- or outward-curling eyelashes, or even an additional row of eyelashes, and all these conditions may require corrective surgery. Glaucoma and PRA (progressive retinal atrophy) are also recognized problems.

AS AN OWNER

These spaniels have extremely high levels of energy, and without enough exercise and stimulation, they may become destructive around the home. Individuals can easily be trained to retrieve balls and to play other fetching games.

Specification

A medium-sized gundog, the English Springer Spaniel has a compact body with a head of proportional size, creating a balanced impression. The coat is of moderate length, with evident feathering. The long, pendulous ears are well covered with hair. The upper lips hang down below the lower jaw line.

RECOGNITION S
North America, Britain and FCI member countries

LIFESPAN
11–13 years

COLOUR
Black and white or liver and white, blue, or liver roan, or tricolour - black or liver and white with tan markings

HEAD
Relatively broad, medium-length skull

EYES
Oval in shape and medium-sized

EARS
Set level with the eye; long and wide, hanging down on the cheeks

CHEST
Deep and of moderate width

TAIL
Carried horizontally or slightly vertically; wagged earnestly by the dog when working

WEIGHT
Dogs 22.5kg (50 lb); bitches 18kg (40 lb)

CHILD FRIENDLINESS

GROOMING

FEEDING

EXERCISE

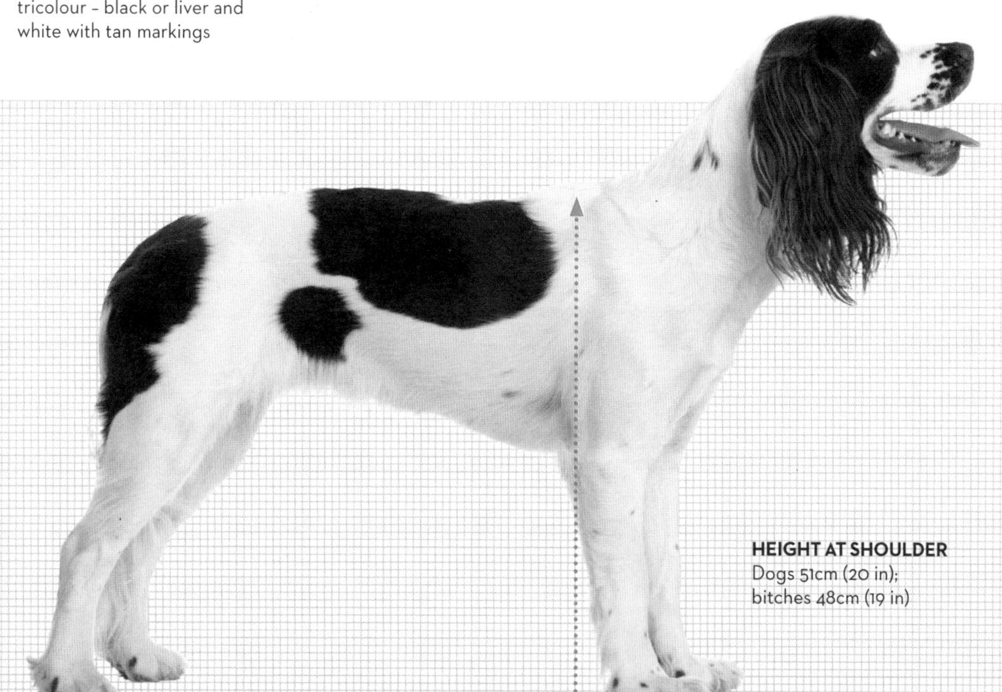

HEIGHT AT SHOULDER
Dogs 51cm (20 in); bitches 48cm (19 in)

SIBERIAN HUSKY

Also known as the Arctic Husky, this is perhaps the best known of the various sled dogs. In spite of their wolf-like appearance, Siberian Huskies are friendly and outgoing, making exuberant companions. The breed was originally developed by the Chukchi people and originated from northeast Asia, before being taken to North America for the first time in the early 1900s.

PERSONALITY

Determined and enthusiastic, the Siberian Husky forms a strong bond with its owner. It does not make a good guard dog, because it is usually open and friendly towards strangers, although older dogs can be more reserved.

HEALTH AND CARE

Siberian Huskies are usually healthy and fit, but they can develop the genetic weakness known as Von Willebrand's disease, affecting the body's blood-clotting mechanism. Medication may be needed after diagnosis, especially for more active dogs.

AS AN OWNER

In common with other sled breeds, the Siberian Husky has plenty of stamina, and it will need a correspondingly large amount of exercise. If you enjoy long country walks, this breed is a good choice as a pet. Unusually among sled dogs, which tend to be prone to having disagreements, Siberian Huskies generally get on well together.

Specification

The Siberian Husky is well-built, with long legs and a medium-length coat, which does not obscure the breed's muscular profile. The outer coat is straight and weather-resistant, with a dense undercoat beneath. Its shaded colouring is highly individual and Siberian Huskies often have striking blue eyes.

RECOGNITION W
North America, Britain and FCI member countries

LIFESPAN
10–12 years

COLOUR
Any colour acceptable

HEAD
Medium-length muzzle, tapering towards the nose

EYES
Almond-shaped; brown, blue, or one of each colour

EARS
Triangular and slightly rounded tips; set high and erect

CHEST
Deep and powerful, but not especially wide

TAIL
Plenty of fur and set just below the topline; carried up, but trailing at rest

WEIGHT
Dogs 20.5–27kg (45–60 lb); bitches 16–22.5kg (35–50 lb)

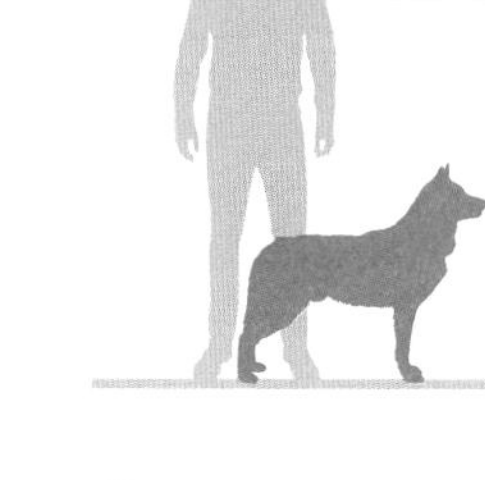

CHILD FRIENDLINESS

GROOMING

FEEDING

EXERCISE

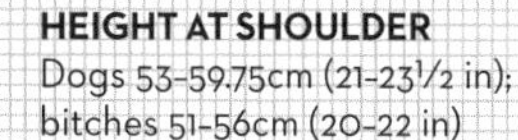

HEIGHT AT SHOULDER
Dogs 53–59.75cm (21–23½ in); bitches 51–56cm (20–22 in)

BEARDED COLLIE

This breed may be descended either from the now-extinct Old Welsh Grey Sheepdog or alternatively from Polish Lowland Sheepdogs, which were known in Scotland as early as the 16th century. Bearded Collies worked as drovers' dogs, overseeing the movement of cattle especially. The breed is so widely-kept today that it is hard to believe that it almost vanished in the 1940s. Since then it has undergone a significant revival, with these sheepdogs becoming popular in North America during the late 1960s.

PERSONALITY

Bearded Collies are active, loyal and affectionate, and possess considerable stamina. The breed thrives with plenty of outdoor exercise and will be an untiring companion on walks.

HEALTH AND CARE

In spite of the limited gene pool from which it was re-created in the 1940s, the breed has proved extremely sound. It needs extensive grooming and a wet, muddy coat is best left to dry and then brushed out. The coat must never be trimmed in show dogs.

AS AN OWNER

Like all herding dogs, the Bearded Collie needs to be kept active if it is not to become bored and destructive.

Specification

The medium-length, profuse coat hangs down the sides of its long lean body. The outer coat is flat and harsh, but the undercoat is softer and provides good insulation. At maturity, black puppies often turn a shaded slate colour, while brown puppies become a shaded dark sand-to-chocolate colour.

RECOGNITION He
North America, Britain and FCI member countries

LIFESPAN
11-13 years

COLOUR
Puppies are brown, fawn, blue or black at birth, sometimes with white markings; most lighten with age

HEAD
Broad, flat skull; strong muzzle and square nose

EYES
Widely spaced, large and expressive

EARS
Medium-length, hanging down the sides of the head

CHEST
Deep, extending at least to the level of the elbows

TAIL
Curved, set and carried low, never above the vertical; can reach the hocks

WEIGHT
18-27kg (40-60 lb)

CHILD FRIENDLINESS

GROOMING

FEEDING

EXERCISE

HEIGHT AT SHOULDER
Dogs 45.5-56cm (21-22 in); bitches 51-53cm (20-21 in)

DALMATIAN

The Dalmatian is one of the most recognizable and distinctive breeds in the world, partly because of the popular film versions of the book *101 Dalmatians*. Nevertheless, it is not a breed that is suited to urban living today. Its precise origins are a mystery, although it is believed to be named after Dalmatia, on the coast of the Adriatic. It has served to protect horse-drawn carriages against highwaymen, by running alongside them. During the 19th century, American fire departments used Dalmatians to help with controlling the horses that pulled fire appliances through busy city streets.

PERSONALITY

Friendly, energetic and tolerant, the Dalmatian is not a shy breed, but its size and exuberance mean that it is not the best choice as a pet in a home alongside young children.

HEALTH AND CARE

This breed may suffer from various hereditary conditions, notably the formation of stones in its urinary tract, which are both painful and potentially dangerous. Deafness is not unknown among puppies. The breed can also develop dermatitis, which may cause the dog to scratch and nibble constantly at its skin.

AS AN OWNER

Dalmatian puppies are white at birth and only develop their spots as they mature. The breed is athletic and makes a great running or jogging companion. Be aware that puppies should not be encouraged to run too far when young.

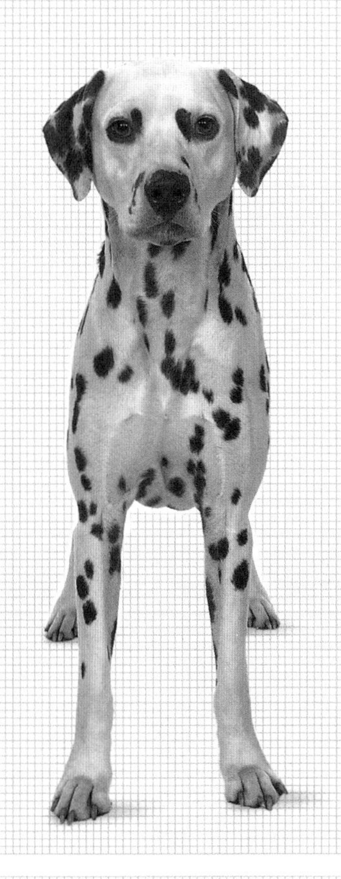

Specification

Dalmatians are characterized by their spotted coat. The patterning is random and enables individual dogs to be told apart easily. In a show dog the spots should be round and must be smaller on the head, legs and tail. They need to be evenly distributed and ideally should not overlap at all.

RECOGNITION NS
North America, Britain and FCI member countries

LIFESPAN
12–14 years

COLOUR
White with liver or black spotting

HEAD
Fair length; tight skin

EYES
Medium-sized, rounded, brown or blue, set deep in the skull

EARS
High, tapering

CHEST
Deep relative to its width

TAIL
Long and tapered

WEIGHT
22.5–25kg (50–55 lb)

CHILD FRIENDLINESS

GROOMING

FEEDING

EXERCISE

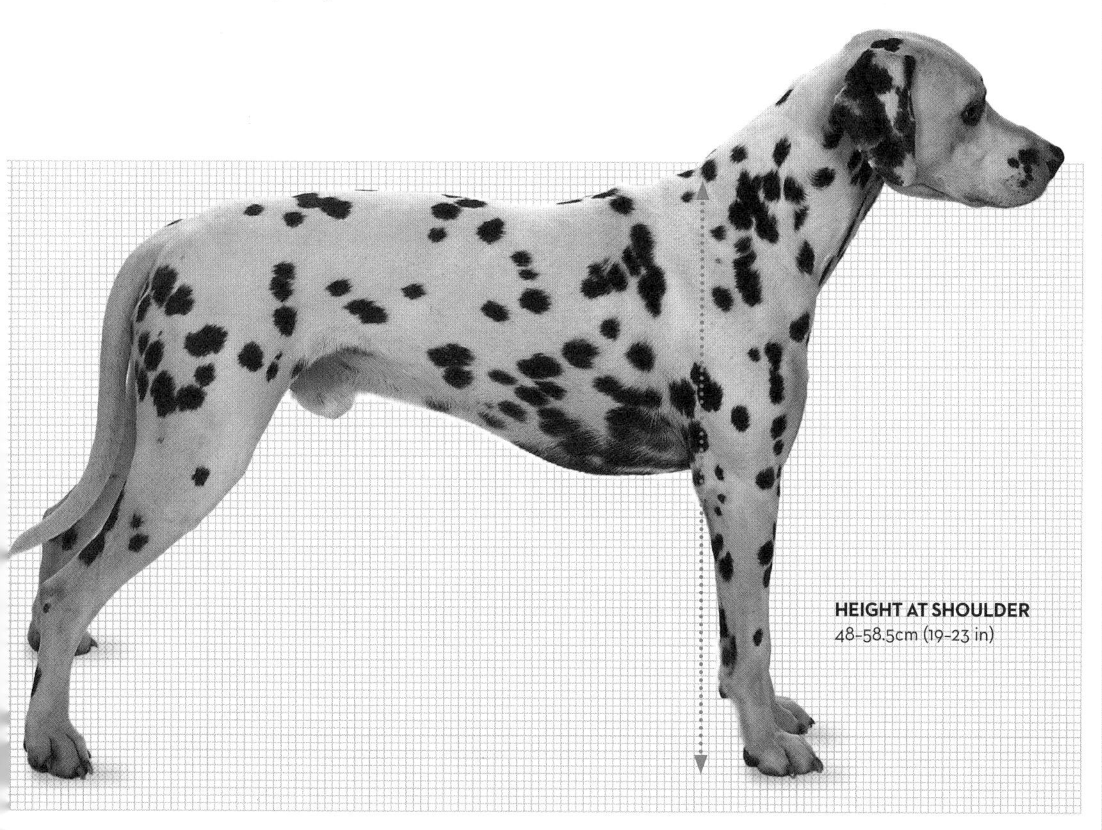

HEIGHT AT SHOULDER
48–58.5cm (19–23 in)

SAMOYED

Under their beautiful and luxuriant coats Samoyeds are tough working dogs, originating from one of the most inhospitable regions on earth. The breed is named after the Samoyede tribe of northern Siberia, who bred these dogs as reindeer herders and also for pulling sleds. The dog's thick fur was made into clothing too. The first Samoyeds seen in the West were brought back to England by fur traders in the late 1800s, and caused a sensation.

PERSONALITY

Intelligent and active, the Samoyed displays great stamina. The breed is naturally friendly and gregarious, but it can also be wayward, particularly if it is bored. When alert, the corners of the mouth turn up, making the dog look as though it is smiling. The independent streak in its nature can make training difficult.

HEALTH AND CARE

Samoyeds are strong dogs, but some congenital circulatory problems are known, and cases of diabetes have been recorded in the breed.

AS AN OWNER

Always prevent a Samoyed from becoming overweight, because obesity is a significant predisposing factor in diabetes. Daily grooming of the thick coat is essential.

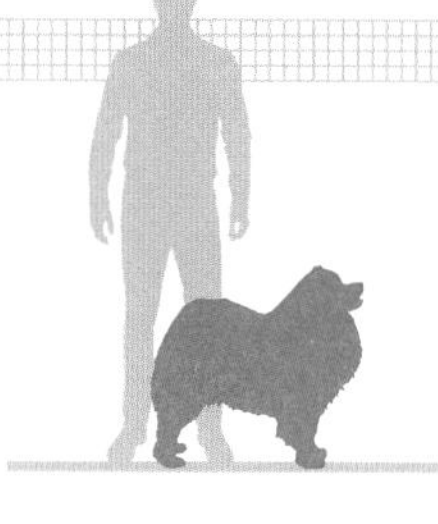

Specification

Pure white is the colour typically associated, but other varieties are accepted. There is a distinctive ruff around the neck, which becomes more profuse during winter. Samoyeds have 'hare' feet – meaning long and narrow – which enable them to walk over snow without sinking into it. Their toes are thickly furred to protect against frostbite.

RECOGNITION W
North America, Britain and FCI member countries

LIFESPAN
10–12 years

COLOUR
Pure white, cream, biscuit, or white and biscuit

HEAD
Broad with a medium-length, tapering muzzle

EYES
Almond-shaped and ideally dark

EARS
Triangular, thick with rounded tips; well-spaced and held erect

CHEST
Deep, reaching down to the elbows

TAIL
Long, with plenty of fur; carried over the back or side and sometimes lowered at rest

WEIGHT
22.5–29.5kg (50–65 lb)

CHILD FRIENDLINESS

GROOMING

FEEDING

EXERCISE

HEIGHT AT SHOULDER
Dogs 53–59.75cm (21–23$\frac{1}{2}$ in); bitches 48–53cm (19–21 in)

NORWEGIAN ELKHOUND

This Nordic hound is a very ancient breed. Archaeological evidence reveals that its ancestors have lived in Norway, where winters can be bitterly cold, for about 7,000 years. The way in which elkhounds and other so-called spitz breeds curl their tails forwards over their backs served as a means of distinguishing them from wolves at a distance.

PERSONALITY

A breed developed to work individually with people, the Norwegian Elkhound makes an intelligent companion. These dogs also possess great energy. They are loyal by nature and make good watchdogs.

HEALTH AND CARE

Plenty of grooming is essential, especially in the spring when elkhounds moult much of their long, thick winter coat. The condition known as progressive retinal atrophy (PRA), causing irreversible damage to the retina and, ultimately, blindness, is a problem for which breeding stock should be screened.

AS AN OWNER

Requires plenty of exercise in the countryside, rather than sedate walks around a park, and a lot of grooming. They are bold dogs – as their name suggests, they were originally bred to hunt elk (better known as moose in North America) – yet their intelligence and versatility ensure that they will learn quickly when trained.

Specification

The breed has upright, triangular ears and a foxlike face, in common with other spitz-type dogs. The dense coat, complete with an insulating under layer, provides excellent protection from the cold. It has a distinctive colour of a deep grey, usually darkest over the saddle area, with a lighter chest and mane of longer fur around the neck. The underparts are silvery, and the muzzle, ears and tip of the tail are black.

North America, Britain and FCI member countries

LIFESPAN
10–12 years

COLOUR
Shades of grey and brownish hues

HEAD
Broad and wedge-shaped, with a tapering muzzle

EYES
Oval, medium-sized and dark brown

EARS
Set high, erect and mobile

CHEST
Deep and relatively broad

TAIL
Tightly curled and set high

WEIGHT
25kg dogs (55 lb); bitches 22kg (48 lb)

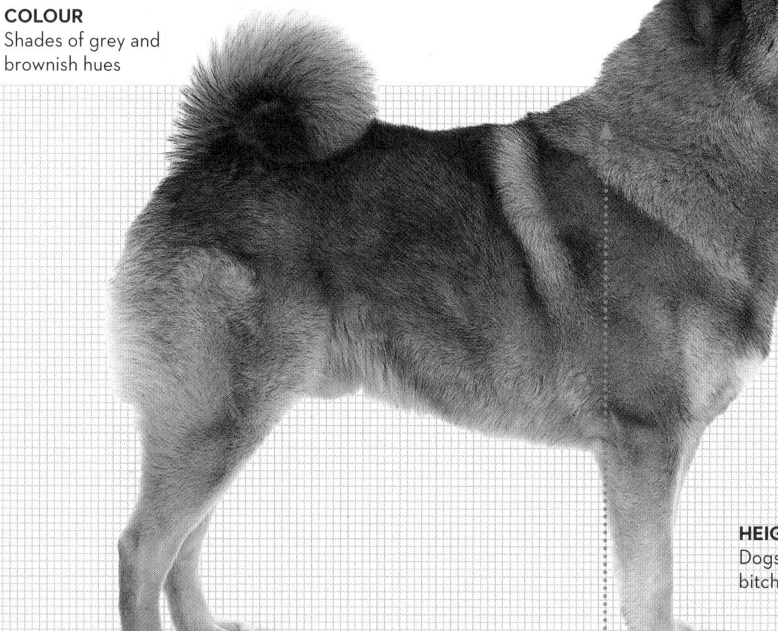

HEIGHT AT SHOULDER
Dogs 54.5cm (21½ in); bitches 49.5cm (19½ in)

MINIATURE BULL TERRIER

Both the Standard and Miniature Bull Terrier are unmistakeable, thanks to their characteristic egg-shaped heads, and possess big personalities. After bull-baiting was banned in 1835 in England, so dog-fighting became popular. Bull Terriers were created by crossings between Bulldogs (see pp. 82–3) and terriers for this purpose, with Dalmatians (see pp. 152–3) also used to add stamina. Miniature Bull Terriers are descended from the runts of Bull Terrier litters, being used for ratting. They stand about 10cm (4 in) shorter.

PERSONALITY

Bull Terriers were ferocious fighters, but are now more gentle and affectionate, although under certain circumstances, they will still act aggressively towards other dogs.

HEALTH AND CARE

The white-coated variety is susceptible to skin cancer, triggered by extensive exposure to hot sun. Avoid exercising these dogs in the middle of the day in hot weather, and rub a canine sunblock on their prominent ears. Pure-white Bull Terriers may also suffer from inherited deafness, a congenital problem linked to their colour.

AS AN OWNER

Even in its Miniature package, the Bull Terrier is very powerful. It must be properly trained from a young age, ideally at puppy classes, were your pet can learn to socialize with other dogs. They then make excellent companions.

Specification

Everything about these terriers emphasizes their strength. Its long head is supported by a thick powerful neck, linked to a broad muscular chest. The body is short and strong, with broad ribs giving a good lung capacity. Although bred in different colours, the white-coated form, with dark spots on the face, is common and reflects their Dalmatian ancestry.

RECOGNITION Te
North America, Britain
and FCI member countries

LIFESPAN
10–12 years

COLOUR
White, fawn, red, black/brindle, plus bi- and tricolour combinations

HEAD
Very distinctive egg-shaped appearance

EYES
Small and dark

EARS
Small and close together; may be held erect

CHEST
Broad and powerful

TAIL
Low-set, short and tapered; carried horizontally

WEIGHT
23.5–28kg (52–62 lb)

CHILD FRIENDLINESS

GROOMING

FEEDING

EXERCISE

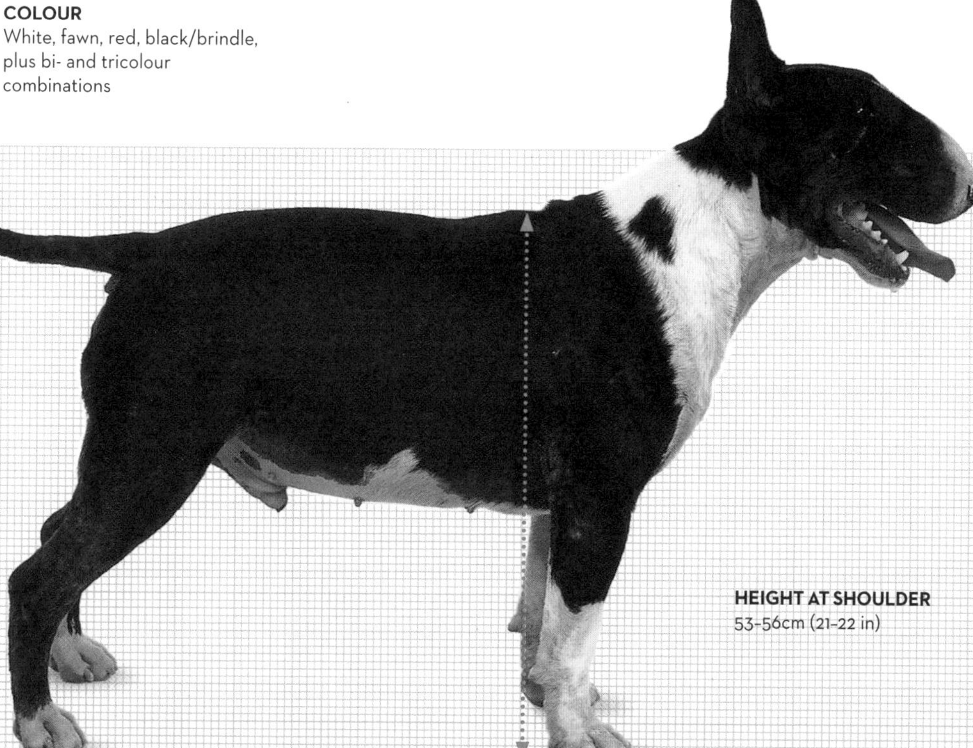

HEIGHT AT SHOULDER
53–56cm (21–22 in)

CHESAPEAKE BAY RETRIEVER

The Chesapeake Bay Retriever is a breed evolved to work in water, named after the area close to Washington, D.C. Since then it has been seen in many other parts of the world, although it still ranks as one of the less common retrievers, probably because it is not well-suited to urban life. The breed's history traces back to 1807, when two Newfoundland puppies (see pp. 226–7) were rescued from a sinking ship off the coast of Maryland. Ultimately, they were mated with local retrieving dogs, laying the foundation for this breed. Other crosses, possibly including Irish Water Spaniels and Otterhounds, also contributed to its subsequent development. Chessies are versatile today, and they are as effective retrieving on land as they are in water.

PERSONALITY

Strong and tireless, these retrievers display immense stamina, able to plunge repeatedly into choppy and often near-freezing coastal waters and retrieve as many as 200 ducks during a day's work.

HEALTH AND CARE

Congenital eye problems such as entropion, which affects the eyelashes and causes these hairs to rub on the surface of the eye, can be an issue. Chessies can also develop progressive retinal atrophy (PRA), so ensure parental stock has been screened. Generally otherwise, Chesapeake Bay Retrievers are extremely hardy dogs.

AS AN OWNER

It is always best to start with a puppy as Chesapeake Bay Retrievers bond early with their immediate family, and it can be difficult to persuade them to transfer their affections at a later stage. They learn readily and are adaptable, as well as being instinctively able to swim well.

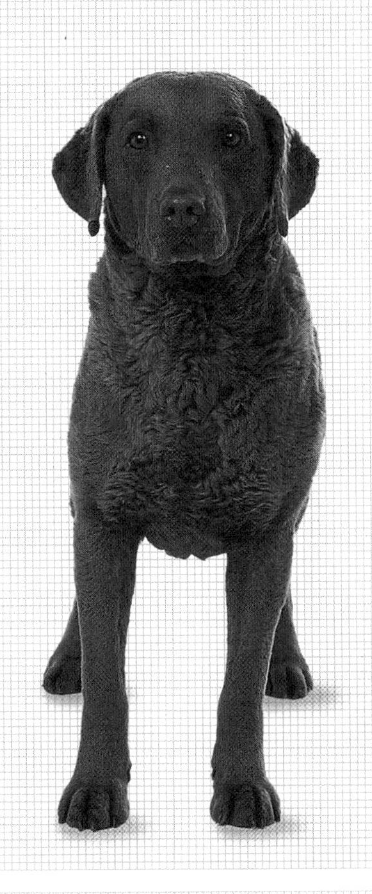

Specification

Signs of its Newfoundland ancestry are still evident in the Chessie's large head and powerful hindquarters. The double-layered coat is a particular feature, providing effective insulation against cold water. The outer layer is also oily and water-repellent, while the insulating woolly undercoat traps warm air close to the skin.

RECOGNITION S
North America, Britain and FCI member countries

LIFESPAN
10–12 years

COLOUR
Shades of brown and sedge

HEAD
Broad, rounded skull

EYES
Widely spaced, relatively large and yellow or amber

EARS
Small, set high, and hang loosely over the sides of the head,

CHEST
Deep, wide and muscular

TAIL
Medium-length; straight or slightly curved, although not over the back

WEIGHT
Dogs 29.5–36kg (65–80 lb); bitches 25–32kg (55–70 lb)

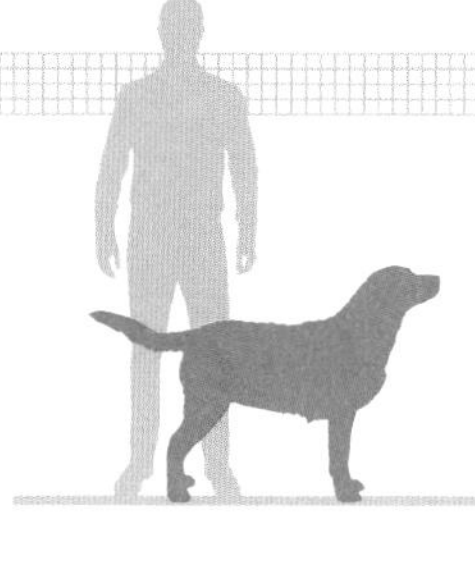

CHILD FRIENDLINESS

GROOMING

FEEDING

EXERCISE

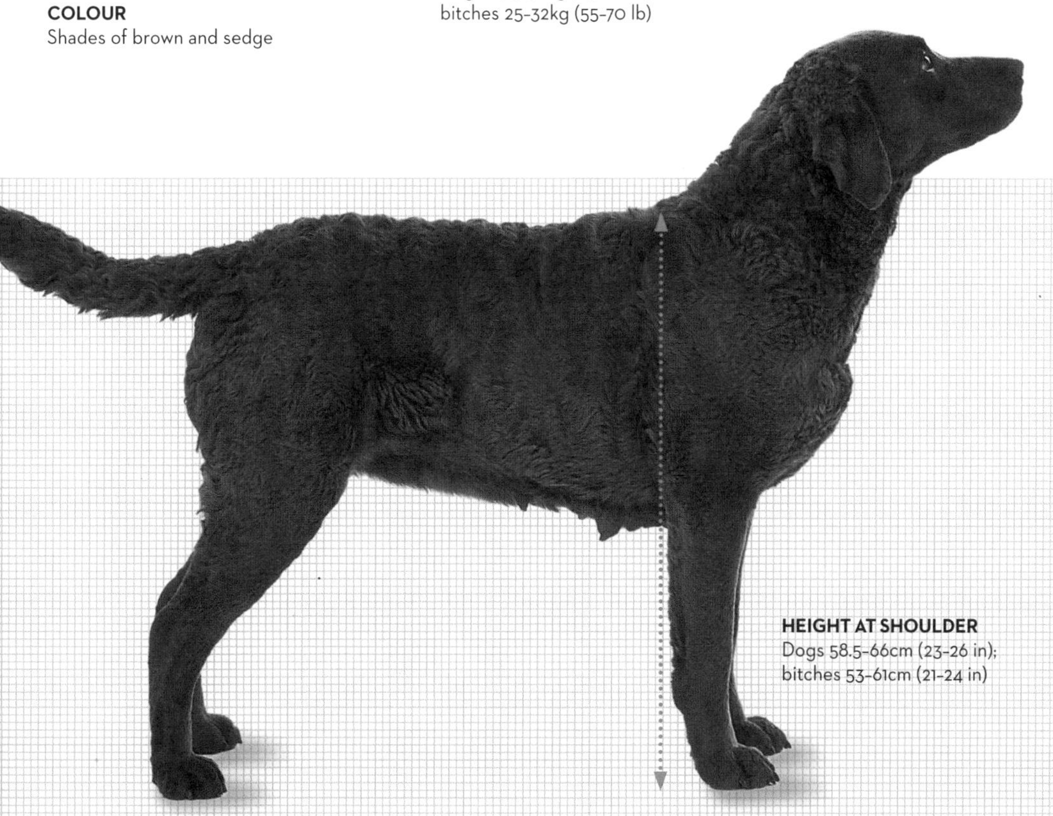

HEIGHT AT SHOULDER
Dogs 58.5–66cm (23–26 in); bitches 53–61cm (21–24 in)

OLD ENGLISH SHEEPDOG

The Old English Sheepdog was created barely 200 years ago and it was never a true sheepdog; it worked instead as a drover's dog, driving both cattle and sheep to market. The breed began life in the southwest of England, being evolved from Bearded Collies (see pp. 150–1) crossed with larger breeds, possibly including the Russian Ovtcharka, which originates in the Ukraine and can measure up to 91cm (3 ft) tall.

PERSONALITY

As good-natured as its appearance suggests, the Old English Sheepdog is a friendly, playful and intelligent breed. Because of its size and natural energy, this dog needs plenty of space if it is to thrive.

HEALTH AND CARE

They may occasionally suffer from cataracts on the eyes, a disorder that is commonly an inherited weakness. Its coat needs regular care to keep it looking its best. In individuals not being shown, it can be clipped back, but this will radically alter the dog's appearance.

AS AN OWNER

Old English Sheepdogs are amiable companions. They are also good guardians, possessing as they do a deep, distinctive bark.

Specification

A profuse covering of hair is a feature of this breed. The coat has a hard texture but tends to fluff out, which gives the dog a decidedly shaggy appearance. This breed moves with little apparent effort, ambling when it walks, and with an elastic movement when trotting or running at full pace. It was called the Bobtail for a time, because of its naturally short tail.

RECOGNITION He
North America, Britain and FCI member countries

LIFESPAN
10–12 years

COLOUR
Grey, grizzled, blue merle or blue, with or without white markings

HEAD
Large and relatively square

EYES
Brown, blue, or a combination of these colours; hidden by hair

EARS
Medium-sized; lie flat against the sides of the head

CHEST
Relatively broad with a deep brisket

TAIL
May be born tail-less

WEIGHT
27–29.5kg (60–65 lb)

CHILD FRIENDLINESS
GROOMING
FEEDING
EXERCISE

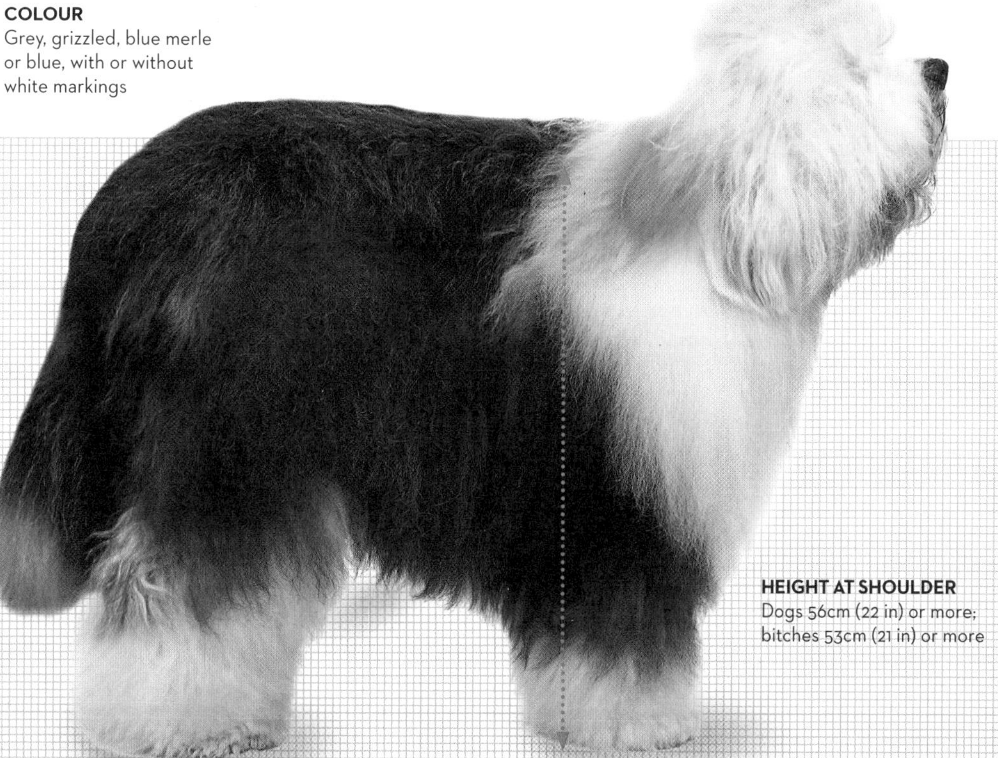

HEIGHT AT SHOULDER
Dogs 56cm (22 in) or more; bitches 53cm (21 in) or more

AIREDALE TERRIER

The largest of the British terrier breeds, the Airedale is an impressive dog with an unmistakable appearance. It is versatile – in addition to being kept as a companion dog, it has undertaken a wide range of working roles, including herding stock, hunting and serving in the Red Cross, seeing active service in World Wars I and II. The Airedale was evolved during the 1840s, close to the Rivers Aire and Wharfe in Yorkshire, England, and its appearance and size suggest that Otterhounds probably contributed to its development. Black and Tan Terriers were also used, as reflected in the breed's colouration. The Airedale reached North America in 1881, where it was used as a gundog.

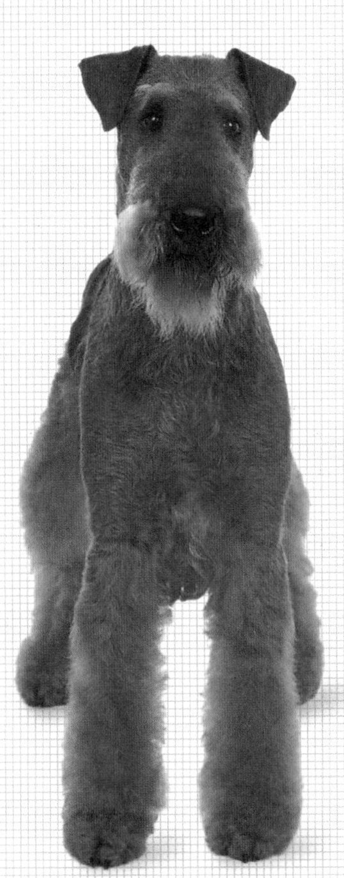

PERSONALITY

Bold and fearless, the Airedale has a strong personality, is energetic and can be playful, but may not always be friendly towards other dogs.

HEALTH AND CARE

Check a puppy's stomach around the umbilicus for lumps or pain. Minor surgery is usually needed to repair this muscle tear. Some young Airedales over 6 months old may be affected by a nervous disorder causes trembling in their hindquarters.

AS AN OWNER

Airedales retain keen hunting instincts and will chase and kill small prey; they should be kept away from pets such as guinea pigs or rabbits. This breed needs a lot of training, which requires patience on the part of their owners.

Specification

The Airedale's face has a distinctive beard, extending down over the jaws. The head and ears must be tan, with the nose being black. The coat is dense, with a hard, wiry texture, but there is a softer undercoat beneath.

RECOGNITION Te
North America, Britain and FCI member countries

LIFESPAN
12–14 years

COLOUR
Tan with black markings forming a 'saddle' on the back

HEAD
Similar length of skull and muzzle

EYES
Small and dark

EARS
Small and V-shaped

CHEST
Deep chest reaches down to the elbows

TAIL
Set high on the back, carried upright

WEIGHT
20–22.5kg (44–50 lb)

CHILD FRIENDLINESS

GROOMING

FEEDING

EXERCISE

HEIGHT AT SHOULDER
Dogs 58.5cm (23 in); bitches 56cm (22 in)

GERMAN SHORTHAIRED POINTER

Lively and enthusiastic by nature, the German Shorthaired Pointer has all the characteristics of a sporting dog, including a requirement for plenty of daily exercise. It will settle well in the home, and makes an alert guardian. Breeds used to develop the German Shorthaired Pointer included the more agile English Pointer (see pp. 202–203) and the Bloodhound (see pp. 196–7), both of which improved its scenting ability, and English Foxhound stock, which contributed greater pace and stamina.

PERSONALITY

The breed is responsive to training. It displays considerable stamina, proving to be a talented retriever, on land and in water.

HEALTH AND CARE

As with many large breeds, hip dysplasia and weaknesses of the elbow joint can be problematic. Epilepsy is also linked with some bloodlines. Do not over-exercise puppies, because this can cause long-term damage to their joints. It has modest grooming needs. Mud should be left to dry and then it can be easily brushed out of the coat.

AS AN OWNER

You need to be an active, outdoor type to take on this breed. German Shorthaired Pointers can be nervous as adults unless they have been carefully socialized early with both other dogs and people, so make sure your puppy is introduced to new experiences in a controlled way while it is still young.

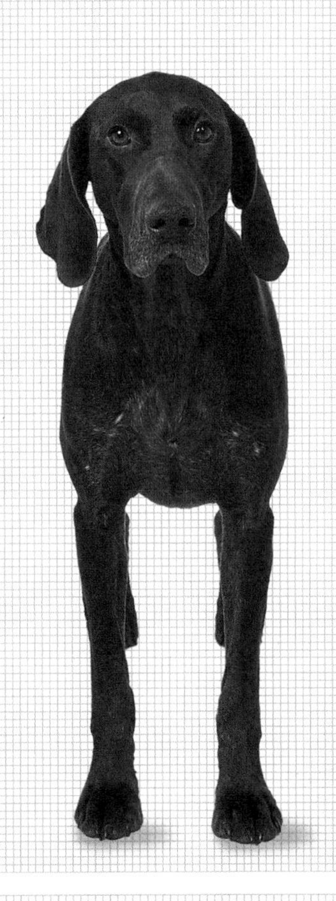

Specification

This is a well-built dog, with a broad skull and strong legs. In bicoloured dogs, the upper parts of the body are coloured, with white areas predominating on the underparts. So-called roan colouring, in which dark and white hairs are merged together, is not unusual.

RECOGNITION S
North America, Britain and FCI member countries

LIFESPAN
12–16 years

COLOUR
Liver, liver-and-white and roan

HEAD
Clean-cut; in proportion to the body

EYES
Medium-sized, lively and intelligent

EARS
Long; should drop level with corners of the mouth

CHEST
Deep and powerful

TAIL
High and firm

WEIGHT
Dogs 25–32 kg (55–70 lb); bitches 20.5–27kg (45–60 lb)

HEIGHT AT SHOULDER
Dogs 58.5–63.5cm (23–25 in); bitches 53–58.5cm (21–23 in)

HUNGARIAN VIZSLA

As a gundog bred to work on the open plains of central Hungary, the Vizsla has been shaped by the environment in which it evolved. It is well adapted to work in the heat, possessing plenty of energy. With its origins dating back more than four centuries, this breed originally served as a pointer, working in combination with falcons, which would seize birds that were flushed after the Vizsla had located them. By the 1800s, however, there were barely a dozen members of this breed alive, but it survived and gained increased international recognition when taken abroad by refugees during World War II.

PERSONALITY

The Vizsla's expression emphasizes its gentle disposition. This is a sensitive breed with a lively, adaptable nature.

HEALTH AND CARE

The blood-clotting disease called haemophilia has been documented, but this is usually a healthy breed. The lack of an undercoat does mean that Vizslas, unlike many gundogs, can feel the cold. This breed may therefore require a coat in bad weather. Its grooming needs are modest, although rare long-haired puppies crop up in litters occasionally.

AS AN OWNER

Vizslas have proved to be one of the most versatile gundog breeds, and even if your pet does not work, it will still enjoy the opportunity to find and retrieve toys, such as balls or flying discs, when out on a walk.

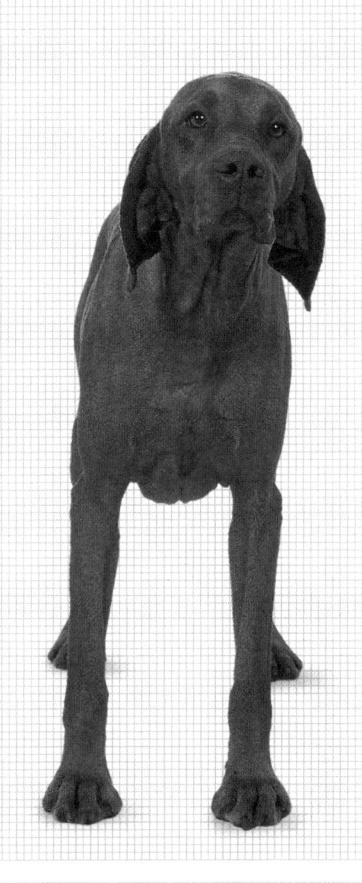

Specification

The sleek, short, shiny golden rust-coloured coat is characteristic, along with a large brown nose and eyes. The hindquarters are powerful, with muscular thighs.

RECOGNITIONS
North America, Britain and FCI member countries

LIFESPAN
11–13 years

COLOUR
Solid golden-rust

HEAD
Muscular, with a square and deep muzzle

EYES
Medium-sized

EARS
Relatively long, set low, hanging against the cheeks

CHEST
Broad, extending down to the elbows

TAIL
Thick at its base and carried almost horizontally

WEIGHT
22–30kg (49–66 lb)

CHILD FRIENDLINESS

GROOMING

FEEDING

EXERCISE

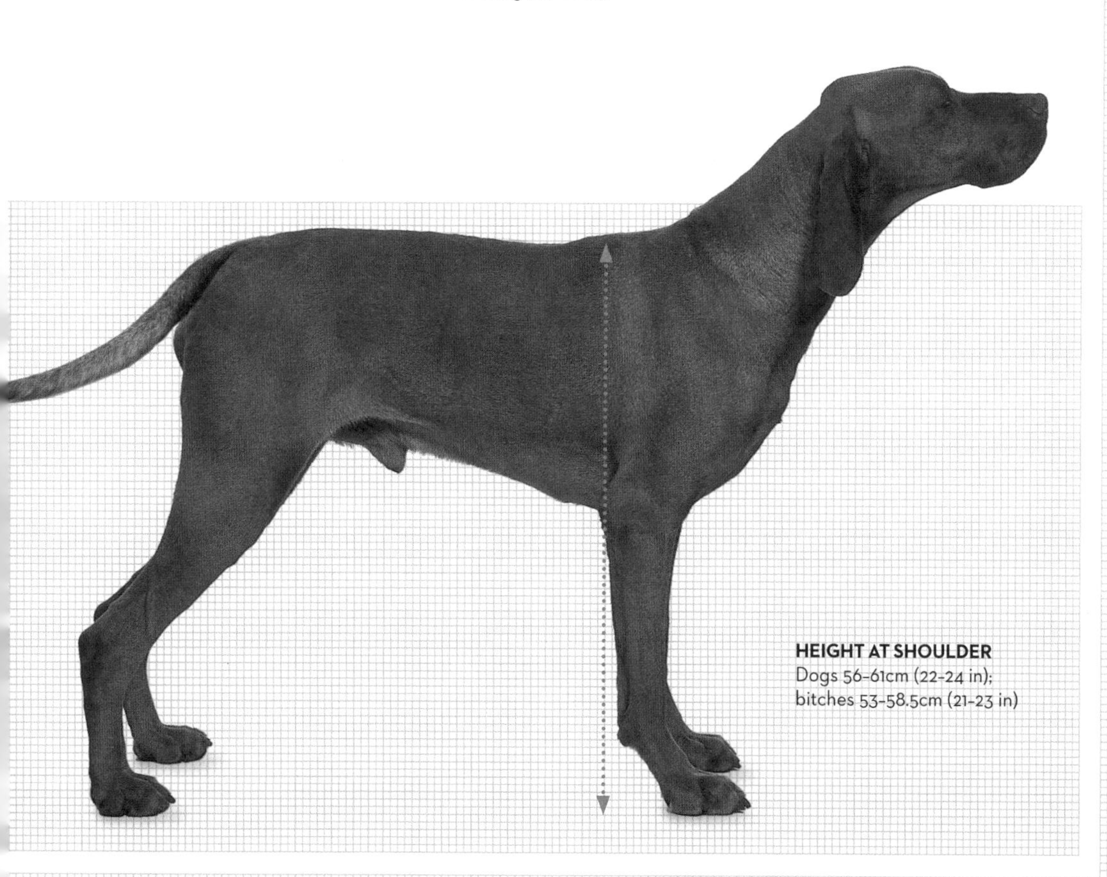

HEIGHT AT SHOULDER
Dogs 56–61cm (22–24 in); bitches 53–58.5cm (21–23 in)

GOLDEN RETRIEVER

Consistently ranking as one of the world's most popular breeds, the Golden Retriever is versatile too, working closely with people in a wide variety of situations today, from guiding the visually impaired to taking an active role in search and rescue operations. The breed's founder was a golden-coloured puppy named Nous born in a litter of black Flat Coated Retrievers (see pp. 180–1) in 1864. Subsequent crossbreeding with the now-extinct Tweed Water Spaniel developed the bloodline, and the distinctive golden colouring became established.

PERSONALITY

Loyal, intelligent and affectionate, with a lively, exuberant side to its nature, the Golden Retriever is universally recognized as the ideal family pet, especially alongside older children.

HEALTH AND CARE

Golden Retrievers may suffer various hereditary problems, particularly some affecting their sight. Cataracts and other conditions affecting the eyelashes and eyelids are not uncommon. Check on screening of breeding stock before buying a puppy.

AS AN OWNER

The Golden Retriever is a reasonably straightforward breed to train, but the puppies tend to be exuberant. If you have no previous experience with dogs, it will be worthwhile attending training classes to master the basics. As adults, these dogs often excel in obedience trials and agility contests.

Specification

Lighter-coloured dogs are preferred in the show ring, and the golden colour of this dog's coat is often paler on the feathering. The coat can be flat or slightly wavy on top, and the dense undercoat acts as an effective barrier against the cold.

RECOGNITION S
North America, Britain and FCI member countries

LIFESPAN
10–12 years

COLOUR
Golden

HEAD
Broad with a straight, slightly tapering muzzle

EYES
Relatively large, set deep and ideally dark brown

EARS
Short and set above and behind the eyes

CHEST
Broad and well-developed

TAIL
Thick and strong, carried either level with the back or slightly upward

WEIGHT
Dogs 29.5–34kg (65–75 lb); bitches 25–29.5kg (55–65 lb)

CHILD FRIENDLINESS

GROOMING

FEEDING

EXERCISE

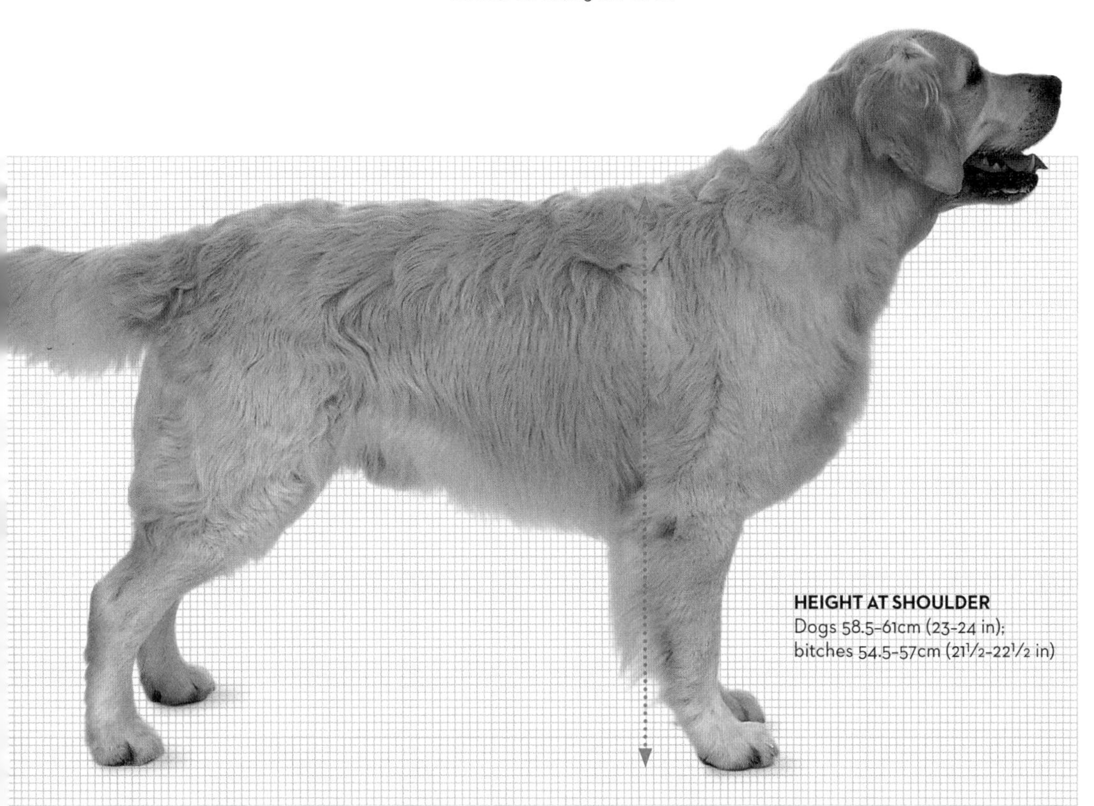

HEIGHT AT SHOULDER
Dogs 58.5–61cm (23–24 in); bitches 54.5–57cm (21½–22½ in)

LABRADOR RETRIEVER

This gundog remains one of the most popular breeds worldwide, developed originally to work closely with people. The Labrador Retriever's ancestors were bred originally during the 1830s by fishermen in Newfoundland. The dogs used their great strength to help with hauling in the nets, and their enthusiasm for swimming is still evident today. The breed was championed by the Earl of Malmesbury, who named it and introduced it to Britain, where it became a highly proficient gundog, both on land and in water.

PERSONALITY

Labradors are lively and enthusiastic dogs. Intelligent and responsive, they are usually straightforward to train and keen to learn. The breed is also versatile, mastering new tasks easily.

HEALTH AND CARE

Labradors are susceptible to hip dysplasia, a problem created by a malformation of the hip joints, causing severe pain. Only purchase puppies from breeding stock that has been screened for this condition. Obesity is a major problem – this is a breed that is extremely fond of food – so provide a balanced diet and plenty of exercise.

AS AN OWNER

With all Labradors, but especially with puppies, be aware of their overwhelming urge to plunge into water – this may occasionally be dangerous, depending on the location. Labradors need training to control their natural exuberance.

Specification

Its tail is distinctive – thick, broad and relatively short – and is sometimes described as an 'otter tail' because it serves as an effective rudder in water. The distinctive glossy coat is short and water-repellent. Black was the original colour, but yellow Labradors are common today, with chocolates also being seen.

RECOGNITIONS
North America, Britain and FCI member countries

LIFESPAN
10–12 years

COLOUR
Yellow, black or chocolate

HEAD
Clean and broad

EYES
Medium-sized and rounded

EARS
Set far back and low on the skull

CHEST
Good balance between width and depth

TAIL
Distinctive otter shape

WEIGHT
Dogs 29.5–36kg (65–80 lb); bitches 25–32kg (55–70 lb)

CHILD FRIENDLINESS

GROOMING

FEEDING

EXERCISE

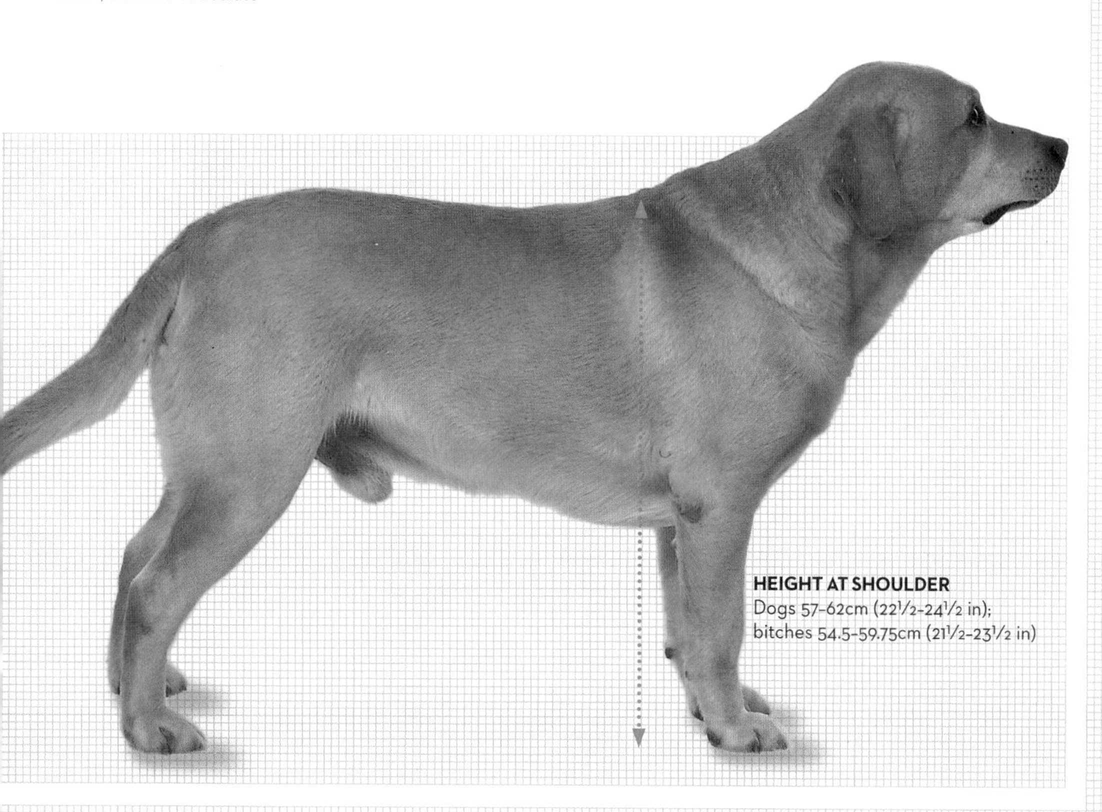

HEIGHT AT SHOULDER
Dogs 57–62cm (22½–24½ in); bitches 54.5–59.75cm (21½–23½ in)

BOXER

There are few dogs with a more playful nature than the Boxer. Its name comes from the way these dogs use their forelegs when leaping up to greet their owners or wrestling good-naturedly with one another. As a boisterous breed, the Boxer is better suited to a home with teenagers than one with young children. Created in Germany in the mid-1800s from crossings between a hunting dog known as the Bullenbeisser and the English Bulldog, the breed was used as a messenger in World War I, before becoming popular as a pet.

PERSONALITY

Highly exuberant, the Boxer is a genuinely good-natured dog, and is remarkably agile as well. Its active nature means that it is not suited to confined living conditions – it needs plenty of space, including access to a garden.

HEALTH AND CARE

Boxer puppies can suffer from various heart and circulatory defects. The breed is also highly susceptible to cancers of many types, so be alert to any possible indicators, such as unexplained swellings on the body or behavioural changes. Early diagnosis can dramatically increase the likelihood of successful treatment.

AS AN OWNER

Few breeds are more enthusiastic than the Boxer, but this behaviour may need to curbed at times particularly in puppies, As a short-nosed breed, it is also important not to exercise Boxers in summer when the sun is at its hottest, because of the risk of heatstroke.

Specification

The Boxer has a distinctive sloping stance, with the hind legs angled back considerably beyond the start of the tail. The patterning of bicoloured Boxers is relatively consistent, with white areas on the muzzle and chest, but no more than a third of the coat should be white.

RECOGNITION W
North America, Britain and FCI member countries

LIFESPAN
9-11 years

COLOUR
Either fawn or brindle, with or without white markings; white Boxers exist, but are not accepted under the breed standard

HEAD
Broad, with the muzzle is higher at the nose than the base

EYES
Large and dark brown

EARS
Set high on the skull, medium-sized and lie over the cheeks

CHEST
Wide, with a deep brisket

TAIL
Broad at the base, often held away behind the body

WEIGHT
27-32kg (66-70 lb)

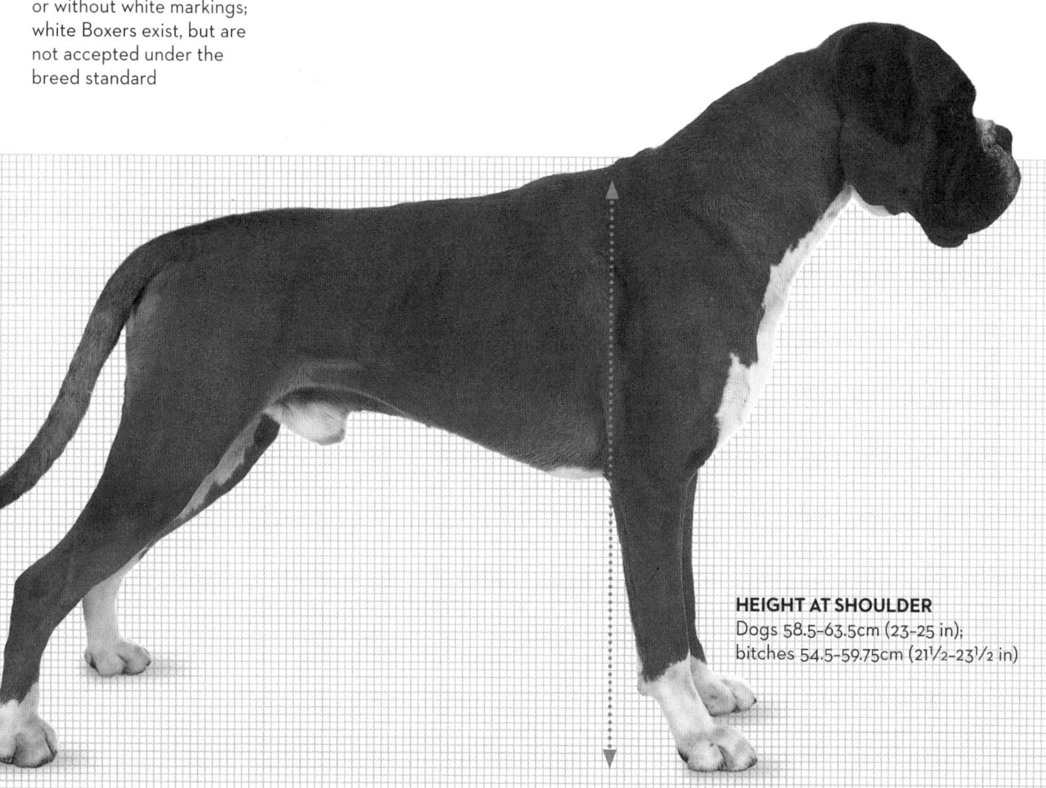

HEIGHT AT SHOULDER
Dogs 58.5-63.5cm (23-25 in); bitches 54.5-59.75cm (21½-23½ in)

TERVUREN

The Tervuren, or Belgian Shepherd Dog, is one of four closely related breeds; another, the Groenendael, from which the Tervuren may have arisen, is known in North America as a Belgian Sheepdog. The Tervuren is classified as a separate breed under its own name however, as are the Belgian Laekenois and the shorthaired Belgian Malinois. These Belgian shepherd dogs probably represent regional variants – the Tervuren is named after the Belgian town where it was first created. It had reached North America by 1918, but the first litter was not born there until 1954.

PERSONALITY

The Tervuren is a hard-working breed that learns quickly: it is successful both in the show ring and in agility competitions, and works both as a guide dog and with the police.

HEALTH AND CARE

Epilepsy can prove a hereditary problem, but an EEG examination can help to diagnose the condition early in life. Grooming is straightforward, in spite of the long dense coat.

AS AN OWNER

The Tervuren is renowned as a guard dog, with an independent, determined nature, and requires sound training. Few breeds display greater loyalty to people they know well.

Specification

The lighter hairs in the Tervuren's coat are tipped with black, a trait which is more prominent in males; the only solid black area is the facial mask, plus the nails.

RECOGNITION He
North America, Britain and FCI member countries

LIFESPAN
10–12 years

COLOUR
Shades of fawn to mahogany, combined with black; a rare silver and black combination exists

HEAD
Skull and muzzle correspond in length, with a reasonably broad muzzle

EYES
Slightly almond-shaped, medium-sized and dark brown

EARS
Set high and erect, with the height the same as the width at the base

CHEST
Deep

TAIL
Kept low at rest; may reach the level of the back when moving

WEIGHT
25–29.5kg (55–65 lb)

CHILD FRIENDLINESS

GROOMING

FEEDING

EXERCISE

HEIGHT AT SHOULDER
Dogs 61–66cm (24–26 in) preferred, but 58.5–67.5cm (23–26½ in) acceptable; bitches 56–61cm (22–24 in) preferred, but 53–62cm (21–24½ in) acceptable

GERMAN WIREHAIRED POINTER

This breed, sometimes called the German Rough-haired Pointer, has been bred on different lines to its shorthaired relative, and is generally less nervous. Although slower, it is a versatile and adaptable gundog. Its precise origins are unclear, but crosses of German Shorthaired Pointers (see pp. 166–7) with other breeds such as the French Griffons, Airedale Terriers (see pp. 164–5) and Pudelpointers probably helped to develop its distinctive appearance.

PERSONALITY

This breed has a reputation for being somewhat aloof with strangers, but it is loyal and affectionate towards people it knows well. Young dogs are enthusiastic and generally learn rapidly.

HEALTH AND CARE

These pointers may suffer from hip dysplasia. Check that any dog you acquire comes from breeding stock that has been screened. Signs of lameness, resulting from underlying arthritis, will otherwise occur. The distinctive coat provides good protection against injuries in the field.

AS AN OWNER

The dense undercoat of the German Wirehaired Pointer is shed in the spring, and much more time must be spent grooming for a short period. Generally, however, coat care is straightforward. If your dog swims or plays in water, the wiry top coat is water-resistant, and after a few quick shakes, it will be dry.

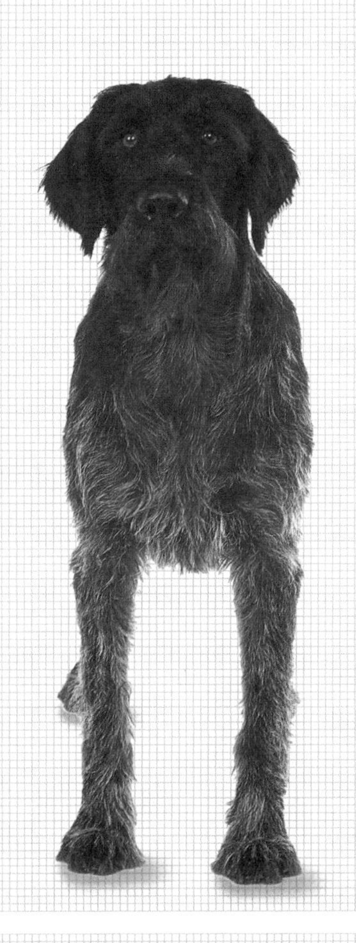

Specification

The coarse beard of these dogs is characteristic, standing out against the shorter hair of the head and ears. The rough coat has a wiry texture, but it lies close to the body and should be no more than 5cm (2 in) long.

RECOGNITIONS
North America, Britain and FCI member countries

LIFESPAN
11–13 years

COLOUR
Liver or liver-and-white combinations

HEAD
Relatively long with a broad skull

EYES
Medium-sized, oval and brown, with protective eyebrows

EARS
Round and hang close to the head

CHEST
Deep

TAIL
Set high and carried either horizontally or above this level when dog is concentrating

WEIGHT
22.5–34kg (50–75 lb)

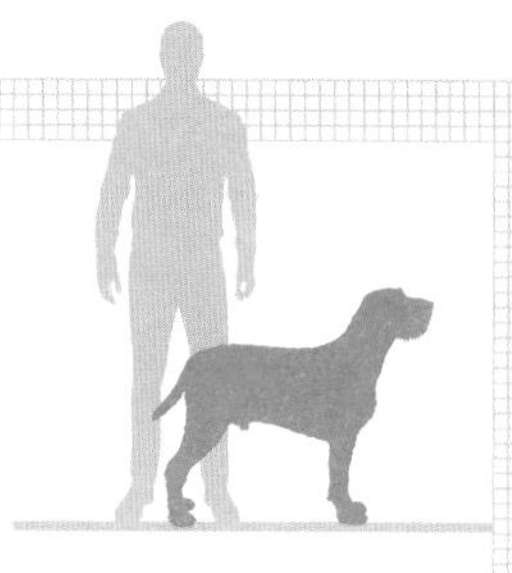

CHILD FRIENDLINESS

GROOMING

FEEDING

EXERCISE

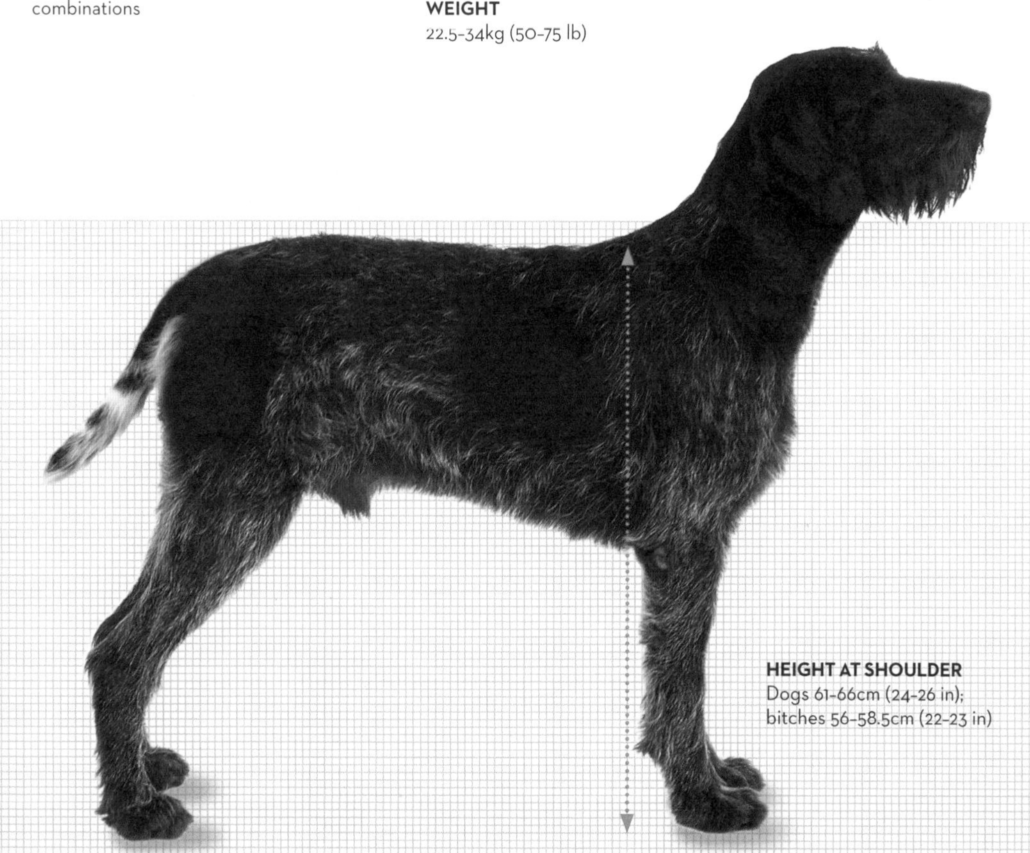

HEIGHT AT SHOULDER
Dogs 61–66cm (24–26 in); bitches 56–58.5cm (22–23 in)

FLAT COATED RETRIEVER

An ancestral form of the Labrador Retriever (see pp. 172–3) from Newfoundland, called the St John's Dog, was crossed with various setters to establish the foundations of this breed. These gundogs were known as Wavy Coated Retrievers for a time, but this setter characteristic gradually disappeared and so their name was changed. Few Flat Coated Retrievers are kept purely as pets today, but they are highly valued for their working ability.

PERSONALITY

Friendly and well-disposed towards people, the Flat Coated Retriever is a breed that is diligent in the field and loyal at home. Its intelligent nature ensures that it is a quick learner, and it proves adaptable and affectionate.

HEALTH AND CARE

The Flat Coated Retriever is generally a healthy breed, although older dogs do seem to be susceptible to bone cancer.

AS AN OWNER

This is not a breed that will thrive in an urban space without having regular exercise each day in rural surroundings. When choosing a puppy, consider that adult dogs and bitches differ from one another in looks more than most breeds – males have a longer and more profuse chest mane.

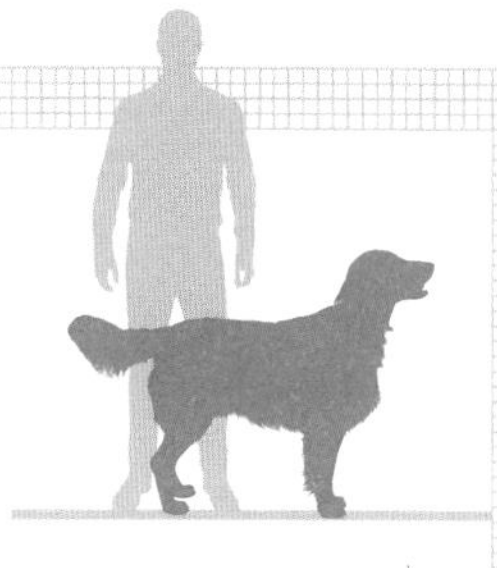

Specification

The most striking feature of the breed is its straight, flat, glossy coat. The legs and tail are well-feathered, and the coat provides good insulation. The head of this dog is large, allowing it to carry game, such as hares or ducks, easily. It works both on land and in water; its muscular hindquarters make it an efficient swimmer.

RECOGNITION S
North America, Britain and FCI member countries

LIFESPAN
11–13 years

COLOUR
Solid black or liver

HEAD
Flat skull and a long, powerful, deep muzzle

EYES
Almond-shaped, brownish, medium sized

EARS
Set relatively high, small yet well-feathered

CHEST
Deep

TAIL
Relatively straight, not curled; not carried significantly above the level of the back

WEIGHT
27–32kg (60–70 lb)

CHILD FRIENDLINESS

GROOMING

FEEDING

EXERCISE

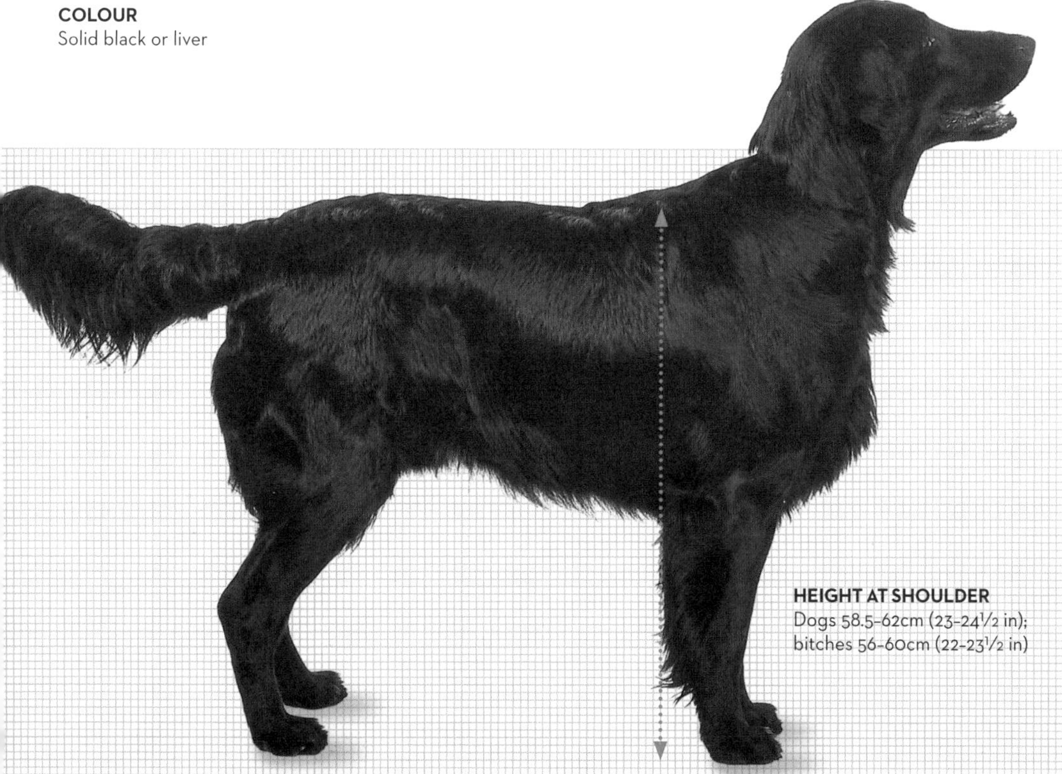

HEIGHT AT SHOULDER
Dogs 58.5–62cm (23–24½ in); bitches 56–60cm (22–23½ in)

ENGLISH SETTER

The description of 'setter' comes from 'setting spaniel', reflecting how these gundogs would sit ('set') quietly once they had detected their quarry, without frightening the birds into flying off. Breeds including the English Springer Spaniel (see pp. 144–5), the Water Spaniel and the Spanish Pointer contributed to the English Setter's development. Various different strains evolved, of which the most famous was created by Sir Edward Laverack in 1825.

PERSONALITY

A breed with great stamina, the English Setter is a good-natured gundog, and makes an excellent companion provided that its need for exercise can be met. It is naturally intelligent and learns quickly.

HEALTH AND CARE

Various hereditary conditions can affect this breed, including hip dysplasia, for which breeding stock should have been screened. A rarer genetic ailment is recessive juvenile amaurotic idiocy. Affected setters show signs of deteriorating vision at 12–15 months, along with some muscle spasms and ultimately seizures occurring.

AS AN OWNER

This is not a breed that will adapt to city life. The English Setter needs space to run, ideally in rural surroundings. If bored, these dogs are likely to become destructive around the home.

Specification

The colouring of this breed is unmistakable: its white coat is overlaid with darker hairs, resulting in flecking known as 'belton'. The silky coat lies flat, with profuse feathering in the chest area and on the back of the legs, the lower underparts, and on the underside of the tail. The pendulous ears are also feathered.

RECOGNITION S
North America, Britain and FCI member countries

LIFESPAN
10–12 years

COLOUR
Lemon with white, liver with white, black with white, or tricolour; plus blue belton with tan markings

HEAD
Long, with a lean shape

EYES
Relatively large and dark brown

EARS
Set low, at or below the level of the eyes, and far back

CHEST
Deep, with brisket reaching to the elbows

TAIL
Extended from the topline; carried straight and level with the back

WEIGHT
25.5–30kg (56–66 lb)

CHILD FRIENDLINESS

GROOMING

FEEDING

EXERCISE

HEIGHT AT SHOULDER
Dogs 63.5cm (25 in); bitches 61cm (24 in)

COLLIE

There are two distinct forms – the longer-haired Rough-Coated form and the Smooth-Coated. These are sometimes classed separately, but they really only differ in terms of their coat length. The Collie's ancestors were probably brought to Britain by the Romans as working sheepdogs more than 2,000 years ago, and later crossed with Borzois (see pp. 228–9), which gave them a more elegant outline and longer legs. A particular favourite of Britain's Queen Victoria, the Rough Collie's profile was then boosted again during the latter half of the 20th century, when the breed starred in the Lassie films.

PERSONALITY

Responsive and affectionate, a Collie can form a very strong bond with its owner, and will have an energetic nature.

HEALTH AND CARE

Collies are prone to various eye conditions. It is important that Merle Collies should not be mated together, because on average, one in four of a resulting litter will be born blind, with small, nonfunctioning eyes.

AS AN OWNER

Rough Collies require plenty of grooming, particularly in spring when the profuse winter coat is shed. These dogs are easy to train, but must be kept apart from sheep, as they retain strong herding instincts.

Specification

The Rough Collie has a long, straight, harsh-textured outer coat, with the sleeker coat of the Smooth Collie highlighting its body outline to a greater extent. Blue merle colouring is common; the mixture of black and white hairs in the coat creates a greyish-blue impression.

RECOGNITION He
North America, Britain and FCI member countries

LIFESPAN
12-14 years

COLOUR
White, blue merle, tricolour or sable and white

HEAD
Lean and wedge-shaped, with a blunt-ended, tapering muzzle

EYES
Obliquely positioned, almond-shaped and medium-sized

EARS
Proportionate to the size of the head; held largely upright when the dog is alert

CHEST
Deep, reaching to the elbows

TAIL
Medium-length and carried low - never over the back

WEIGHT
Dogs 27-32kg (60-70 lb); bitches 22.5-29.5kg (50-65 lb)

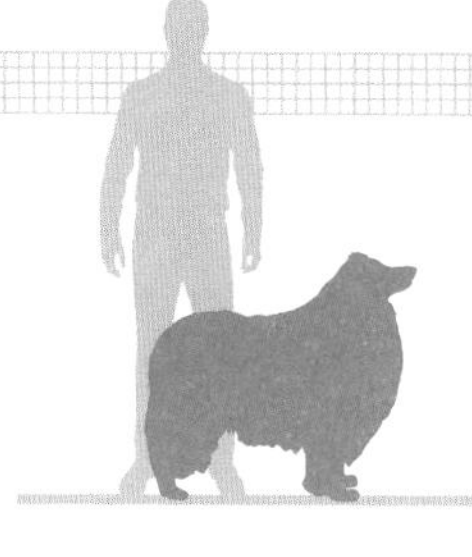

CHILD FRIENDLINESS

GROOMING

FEEDING

EXERCISE

HEIGHT AT SHOULDER
Dogs 61-66cm (24-26 in); bitches 56-61cm (22-24 in)

GERMAN SHEPHERD DOG

The German Shepherd Dog was also known as the Alsatian for a period, because of anti-German sentiment over World Wars I and II. It was developed using a number of localized herding dogs, which tended to have straighter backs and taller, squarer profiles. The familiar form of the breed today is the result of a dedicated breeding programme started in the 1890s by a Captain Max von Stephanitz.

PERSONALITY

Slightly reserved but confident, the German Shepherd has proved to be a loyal and highly intelligent breed that learns quickly. It is often used by the police and military, but can equally serve as a guide dog, thanks to its dependable and responsive nature.

HEALTH AND CARE

Beware of hip dysplasia. A screening programme for breeding stock is now commonplace, but always check the parents have been formally screened, before buying a puppy. Any prolonged digestive upsets could indicate pancreatic problems.

AS AN OWNER

Only contemplate acquiring a German Shepherd if you are prepared to invest adequate time in training. Few breeds are more responsive. Regular grooming of the double-layered, dense coat is essential, particularly during the twice-yearly moult.

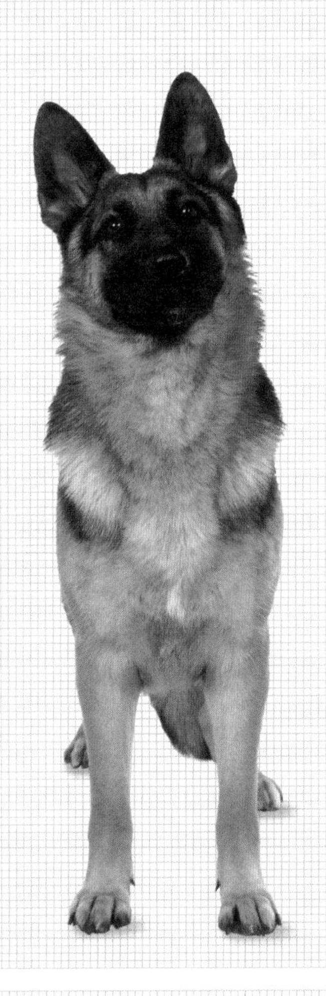

Specification

The graceful downward curve of the German Shepherd's top line should be continuous when seen in profile, extending from the pointed ear tips right down to the end of the tail. Its gait should be in the form of an easy trotting movement that covers a considerable amount of ground with every stride.

RECOGNITION He
North America, Britain and FCI member countries

LIFESPAN
10–12 years

COLOUR
All colours (there areshow restrictions on the white form)

HEAD
Skull extends into the long wedge-shaped muzzle with no stop

EYES
Almond-shaped, set slightly obliquely

EARS
Erect, moderately pointed and balanced in size on the skull

CHEST
Deep

TAIL
Never curled forwards beyond a vertical point; carried in a sabre-like position at rest

WEIGHT
34–43kg (75–95 lb)

CHILD FRIENDLINESS

GROOMING

FEEDING

EXERCISE

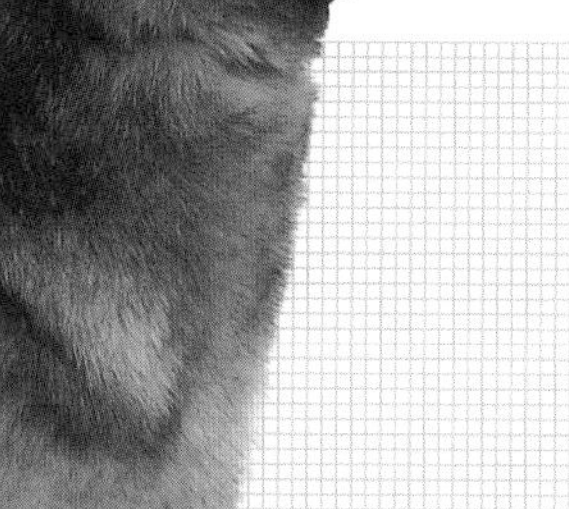

HEIGHT AT SHOULDER
Dogs 61–66cm (24–26 in); bitches 56–61cm (22–24 in)

WEIMARANER

These elegant gundogs, with their distinctive silvery grey coats, were so highly prized originally that they could only be kept by the aristocracy. Weimaraners have now acquired a strong international following. This breed was created at the court of Grand Duke Karl August of Weimar in about 1810. He sought to breed the ultimate hunting companion, using a combination of German pointers, plus Bloodhounds (see pp. 196–7) for their scenting skills and French hounds contributing pace and extra stamina. The Weimaraner's original quarry was large dangerous mammals, such as bears and wild boar, but as they became scarcer in Europe, so the breed evolved into a bird dog.

PERSONALITY

The Weimaraner is a gundog that was developed also as a companion breed. It is loyal, friendly and adaptable, whether working on land or retrieving in water.

HEALTH AND CARE

Check for any swelling around a puppy's umbilicus, indicating a possible hernia here. This breed is also prone to skin problems, often reflected by repeated scratching or biting of the skin. Parasites or even food intolerances may be responsible.

AS AN OWNER

A Weimaraner needs reasonably spacious surroundings with a secure garden. Puppies respond well to training. When grooming, use a hound glove to impart a good gloss.

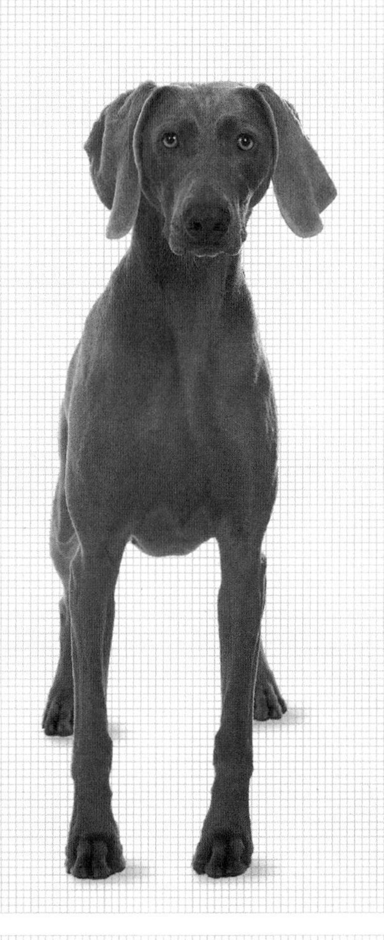

Specification

It is not just the Weimaraner's sleek glossy coat that is an unusual colour – the eyes are distinctive too, ranging from amber to grey or blue-grey. The forelegs are strong and straight, with the hindquarters being well-muscled, creating an apparently effortless gait.

RECOGNITIONS
North America, Britain and FCI member countries

LIFESPAN
11–13 years

COLOUR
From silvery grey to a darker mouse grey

HEAD
Relatively long with a muzzle as long as the skull

EYES
Well-spaced

EARS
Set high, slightly folded and long

CHEST
Deep and well-muscled

TAIL
Carried slightly vertically, conveying confidence

WEIGHT
32–39kg (70–86 lb)

GROOMING

FEEDING

EXERCISE

HEIGHT AT SHOULDER
Dogs 63.5–68.5cm (25–27 in); bitches 53–63.5cm (23–25 in)

ROTTWEILER

The Rottweiler has an intelligent and responsive nature. These are powerful dogs, however, and need to be properly controlled at all times. This breed displays strong territorial instincts and often will not take well to newcomers on your premises, so careful introductions are required. Named after the town of Rottweil in southwest Germany, the Rottweiler was probably developed by crossbreeding between ancient mastiff stock and sheepdogs native to the region.

PERSONALITY

Strong-minded and confident, the Rottweiler will instinctively seek to defend its territory, especially if challenged. These dogs respond positively to training, displaying great devotion to those they know well.

HEALTH AND CARE

Rottweilers can develop entropion, a condition in which the eyelids turn inwards, rubbing on the eyeball and causing irritation. They are also vulnerable to diabetes mellitus. Symptoms include weight loss, constant hunger and thirst.

AS AN OWNER

The Rottweiler has an instinctively dominant nature and firm training from puppyhood is essential, particularly because these are immensely strong dogs. This is a breed that will benefit from socialization at puppy-training classes too. A responsive Rottweiler will then be an excellent companion dog. It has minimal grooming needs.

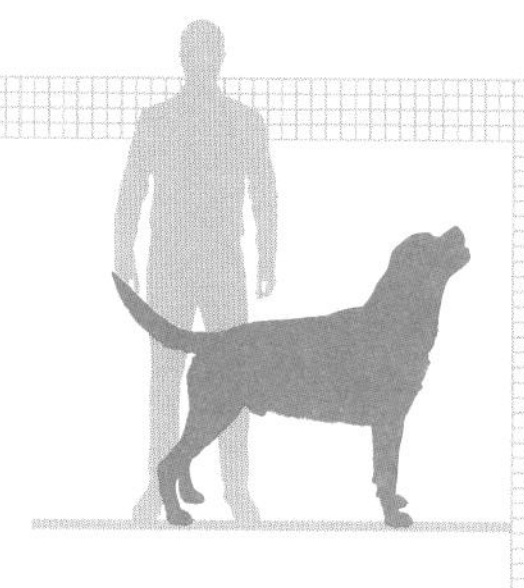

Specification

The breed has a short, straight, flat coat, which is predominantly black, with distinctive rust markings. These should be present over each eye, on the cheeks and the sides of the muzzle, on each side of the chest and on the lower part of the front legs, as well as covering much of the lower hind legs.

RECOGNITION W
North America, Britain and FCI member countries

LIFESPAN
10–12 years

COLOUR
Black with rust markings

HEAD
Medium-length, broad between the ears, with a powerful muzzle

EYES
Almond-shaped and dark brown

EARS
Triangular and medium-sized; hang forwards

CHEST
Deep, broad and muscular

TAIL
Set creates an impression of lengthening the topline; can be carried above the horizontal

WEIGHT
41–50kg (90–110 lb)

CHILD FRIENDLINESS

GROOMING

FEEDING

EXERCISE

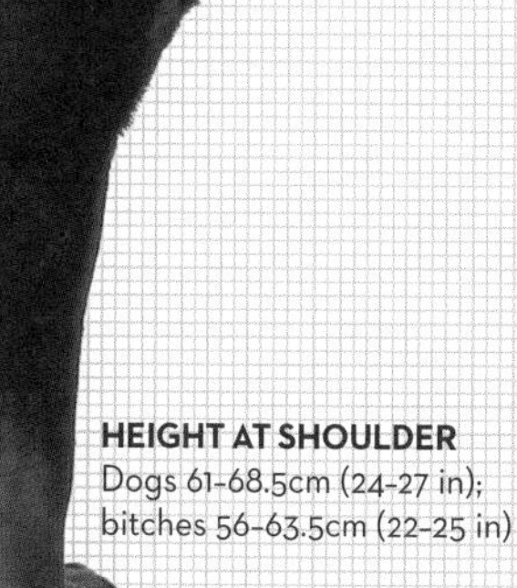

HEIGHT AT SHOULDER
Dogs 61–68.5cm (24–27 in); bitches 56–63.5cm (22–25 in)

GORDON SETTER

This breed is also known as the Black and Tan Setter, thanks to its colouration. It is named after the Duke of Richmond and Gordon, who developed it in the 1820s on his Scottish estate in Banffshire. He wanted a setter that could hunt game birds, such as partridge and woodcock, under more testing conditions than the English Setter (see pp. 182–3). He therefore created a breed that was larger and more powerful albeit not as fast. It was popular back in the late 1800s, but is seen less often today. Its decline in popularity is a pity because it is both an attractive and an affectionate breed.

PERSONALITY

Hard-working and strongly devoted to family members, the Gordon Setter is far less inclined to accept visitors, serving as an alert watchdog. These dogs are intelligent and confident, and they also have admirable recall when it comes to training and for remembering people and places.

HEALTH AND CARE

A hereditary illness occasionally associated with this breed is progressive retinal atrophy (PRA), an incurable condition that ultimately leads to blindness. Breeding stock should be screened for this weakness.

AS AN OWNER

This breed usually has more energy than its relatives, so be prepared for long daily walks if you choose this setter.

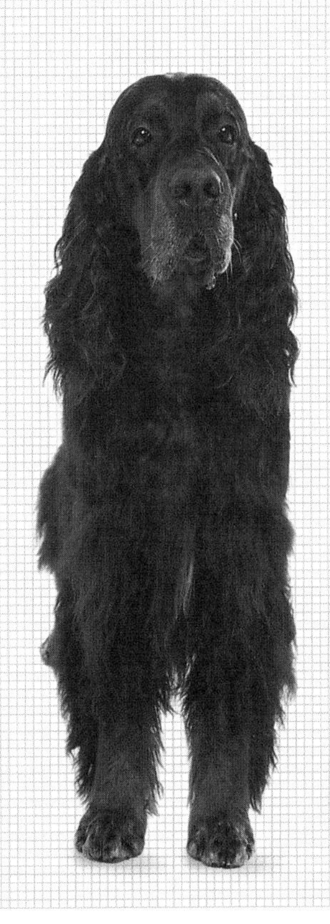

Specification

Black and tan colouration is now the only colour combination, although originally black- or red-and-white bicolours existed, as well as black, tan, and white tricolours. The Gordon Setter's glossy fur is often slightly wavy and its large nostrils aid its scenting ability.

RECOGNITIONS
North America, Britain and FCI member countries

LIFESPAN
11–13 years

COLOUR
Black with tan markings

HEAD
Deep, with a long but not pointed muzzle

EYES
Oval and dark brown

EARS
Large, folded and set low; carried close to the head

CHEST
Deep, but not particularly broad

TAIL
Short and tapered; carried horizontally or nearly horizontally

WEIGHT
Dogs 25–36kg (55–80 lb); bitches 20.5–32kg (45–70 lb)

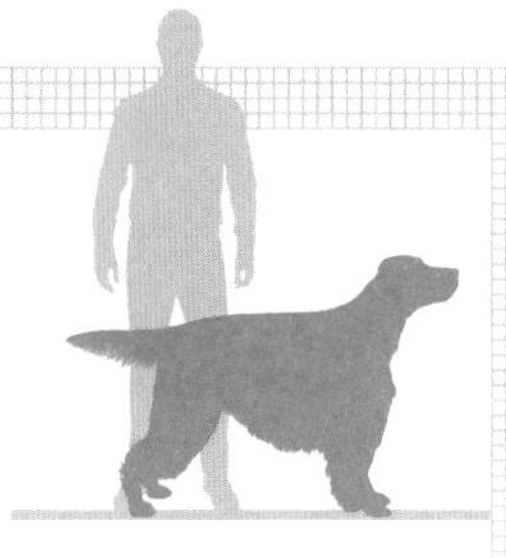

CHILD FRIENDLINESS

GROOMING

FEEDING

EXERCISE

HEIGHT AT SHOULDER
Dogs 61–68.5cm (24–27 in); bitches 58.5–66cm (23–26 in)

ALASKAN MALAMUTE

These powerful sled dogs tend to have dominant natures, and are best-suited to experienced owners. Like other breeds from the far north, Alaskan Malamutes usually communicate by howling rather than by barking. Originally bred by the Mahlemuts, an Inuit tribe from northwest Alaska, Malamutes have a long history. Long before the advent of motorized vehicles, they provided a vital means of transport in the frozen wastelands there. Recently, as sled racing developed into a sport, so the breed grew in international popularity.

PERSONALITY

Strong-willed and determined, Alaskan Malamutes work in teams, but they form a distinct hierarchy within the group, as occurs in a wolf pack. The result is that males in particular can often be aggressive towards other dogs.

HEALTH AND CARE

This breed can suffer from various inherited conditions, including a form of hereditary dwarfism, which causes puppies to be born with short legs. Other genetic problems include a susceptibility to kidney disease and hemeralopia, an eye condition that results in dogs being blinded by bright light.

AS AN OWNER

The immense power of these dogs means that they must be taught to respond to you from an early age. Male dogs can be more assertive and instinctively dominant by nature, and this also needs to be borne in mind, with neutering being recommended.

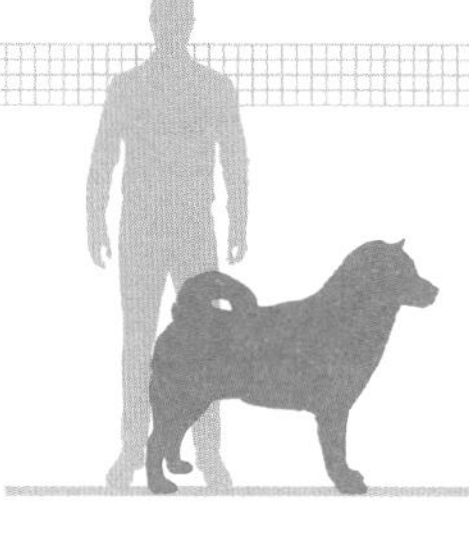

Specification

The Alaskan Malamute has a powerful body with strong shoulders, for pulling the sleds, and a deep chest with plenty of lung capacity, along with sturdy legs and tight, deep paws. The breed is also protected against the elements by its woolly undercoat, which can be up to 5cm (2 in) long. The coarser outer coat is thick and longest on the back of the body.

RECOGNITION W
North America, Britain and FCI member countries

LIFESPAN
10–12 years

COLOUR
Solid white, or a shaded appearance ranging from grey to black and from sable to red

HEAD
Broad with a large, powerful muzzle

EYES
Medium-sized, brown and almond-shaped

EARS
Well-spaced, rounded at the tips and usually erect

CHEST
Deep and powerful

TAIL
Resembles a plume, with plenty of fur, and carried forwards at rest

WEIGHT
Dogs 38.5kg (85 lb); bitches 34kg (75 lb)

CHILD FRIENDLINESS

GROOMING

FEEDING

EXERCISE

HEIGHT AT SHOULDER
Dogs 63.5cm (25 in); bitches 58.5cm (23 in)

BLOODHOUND

The Bloodhound's name simply reflects the fact that it has a long lineage, with these being gentle hounds that develop a very close rapport with their owners. They are descended from the ancient St Hubert Hound, which was first recorded around AD 600. The dog's remarkable tracking abilities were first employed to locate injured stags, but were later used to track people.

PERSONALITY

Lively and friendly, but with a determined nature, the Bloodhound is unlikely to be thrown off the scent, even if this is days old. They have been recorded following cold trails for up to 220km (138 miles).

HEALTH AND CARE

The skin folds on the head can occasionally become infected, with bathing of this area then being necessary. The eyelids can also be a cause for concern. This breed suffers both from ectropion, when the eyelids extend away from the eyes, and entropion, in which the eyelashes rub on the surface of the eye. Surgical correction is likely to be necessary for both.

AS AN OWNER

Be prepared for long walks every day with this breed, which makes an affectionate companion. The Bloodhound has a wonderful baying call, which it will use to indicate its presence to you, particularly in woodland. Grooming needs are minimal.

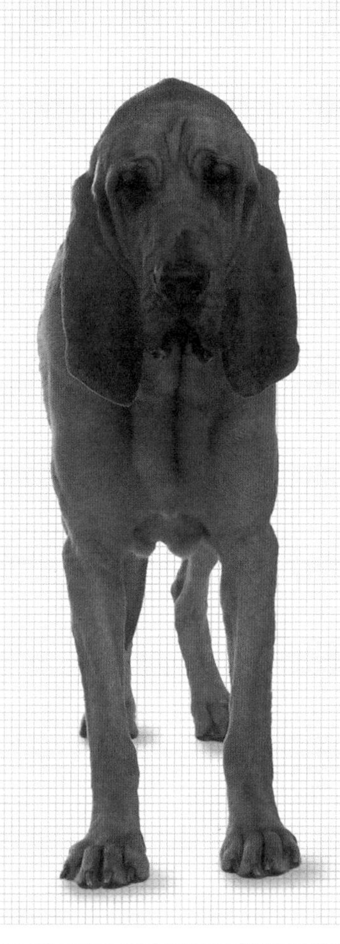

Specification

Often solid tan with a black area forming a 'saddle' on the back. The ears are very long and pendulous while the nose is distinctively broad, assisting scent-detection.

RECOGNITION H
North America, Britain
and FCI member countries

LIFESPAN
10–12 years

COLOUR
Black and tan; liver and tan; red

HEAD
Long and yet relatively narrow, with folds of loose skin

EYES
Well-sunken; hazel often preferred

EARS
Set low, forming folds on the sides of the head

CHEST
Extends down low between the forelegs, forming a keel

TAIL
Carried upright and curved in a scimitar fashion

WEIGHT
Dogs 41–50kg (90–110 lb);
bitches 36–45.5kg (80–100 lb)

GROOMING

FEEDING

EXERCISE

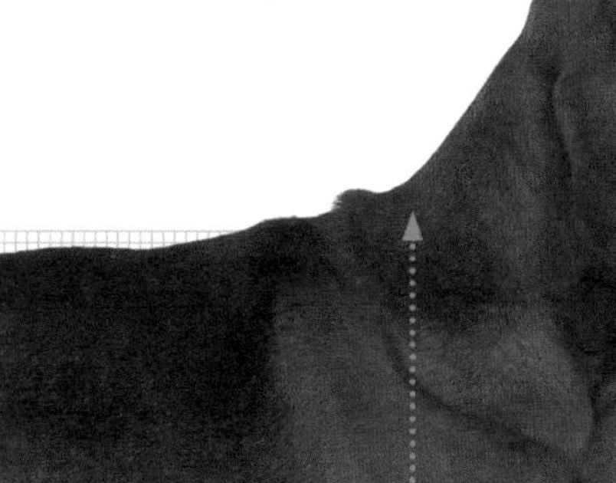

HEIGHT AT SHOULDER
Dogs 63.5–68.5cm (25–27 in);
bitches 58.5–63.5cm (23–25 in)

HOVAWART

The name of this breed is pronounced 'hoffavart', meaning 'farm guardian.' First recorded in the 13th century, such dogs used to be a common sight working on farms in Germany. By the early 20th century, however, the Hovawart was virtually extinct. It was recreated with a combination of breeds such as the Leonberger and the German Shepherd Dog (see pp. 186–7), and is becoming popular as a family companion, proving a loyal and alert companion.

PERSONALITY

Affectionate and loyal, today's Hovawart tends to be kept more as a pet than used as a working breed, although it retains its protective instincts. It can be trained easily, and it has a surprisingly playful side, often settling in well with children.

HEALTH AND CARE

Unexplained weight gain and a loss of energy may indicate hyperthyroidism – an underactive thyroid gland. This condition, to which the breed is susceptible, can usually be treated successfully. Regular grooming is also important.

AS AN OWNER

This is an active breed that needs plenty of exercise. Do not be tempted by a Hovawart if you are living in a city environment where there is relatively little space to exercise your pet. These dogs will thrive, however, if they can have regular country walks, and may display surprisingly good scenting skills.

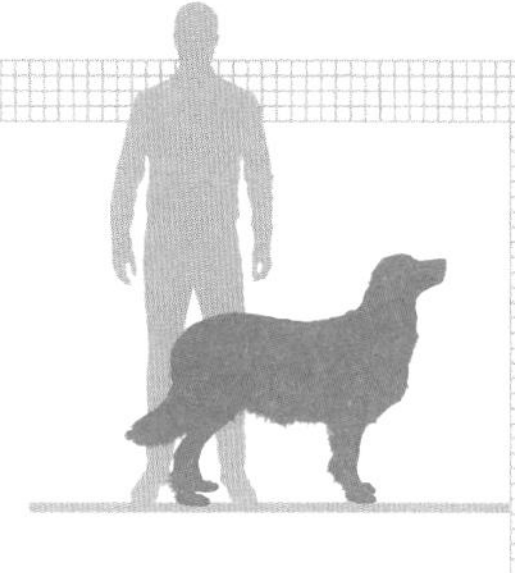

Specification

The Hovawart looks more like a gundog, being reminiscent of the Flat Coated Retriever (see pp. 180–181). The topcoat is long and the undercoat may be slightly wavy. There is distinctive feathering on the backs of the front legs.

RECOGNITION W
FCI member countries

LIFESPAN
10-12 years

COLOUR
Black, blond, or black and gold

HEAD
Strong, with a relatively broad muzzle

EYES
Oval-shaped, medium-sized, and dark

EARS
Pendulous, triangular, and set towards the back of the head

CHEST
Deep and powerful

TAIL
Set just below the topline, extends to below the hocks, well-feathered and carried up when alert

WEIGHT
Dogs 30-40kg (66-88 lb); bitches 25-35kg (55-77 lb)

CHILD FRIENDLINESS

GROOMING

FEEDING

EXERCISE

HEIGHT AT SHOULDER
Dogs 63.5-70cm (25-27½ in); bitches 58.5-64.75cm (23-25½ in)

BERNESE MOUNTAIN DOG

The Bernese Mountain Dog has been associated with the area of Berne in the Swiss Alps for centuries. It is the result of crossings between flock guardians and mastiff stock. These dogs proved versatile working companions, best-known for pulling carts carrying agricultural produce, such as milk, for the cheese-making industry there. By the 1890s, however, numbers had fallen, but thanks to a few dedicated enthusiasts, this attractive breed has survived.

PERSONALITY

A patient, tolerant breed, the Bernese Mountain Dog makes an excellent family pet. It gets along well with children and often displays considerable patience with them. It is also an easy breed to train.

HEALTH AND CARE

Occasionally, puppies are born with blue eyes, which is a serious show fault, although such individuals still make ideal companions. Another occasional genetic problem evident in puppies is a cleft palate or lip.

AS AN OWNER

Individuals need plenty of exercise. The natural gait is a trot rather than a walk. Some owners still train their dogs to pull carts, in the traditional manner. Grooming is relatively straightforward, in spite of the length of the coat, and serves to maintain its characteristic glossy sheen.

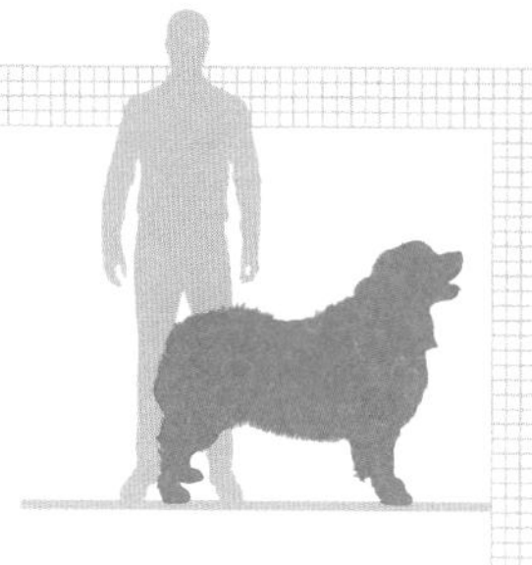

Specification

The Bernese Mountain Dog can be distinguished from related Swiss breeds by its long and slightly wavy coat and larger size. The markings of these dogs are distinctive, with rust-coloured patches over the eyes and a white blaze extending down from the forehead to the muzzle. The feet and the chest must also be white, forming a broad and cross-like chest marking.

RECOGNITION W
North America, Britain and FCI member countries

LIFESPAN
10–12 years

COLOUR
Tricoloured: white, black and rust in a set pattern

HEAD
Broad, with a strong, straight muzzle

EYES
Oval and dark brown

EARS
Triangular and rounded at the tips

CHEST
Deep, reaching to the elbows

TAIL
Bushy; carried low and never over the back

WEIGHT
38.5–41kg (85–90 lb)

CHILD FRIENDLINESS

GROOMING

FEEDING

EXERCISE

HEIGHT AT SHOULDER
Dogs 63.5–70cm (25–27½ in); bitches 58.5–66cm (23–26 in)

POINTER

These gundogs communicate when working not by barking, but by freezing (or 'pointing') in a characteristic stance when they detect their game, revealing exactly where the birds are hidden in undergrowth. Almost certainly, the English Pointer is descended from one of the European pointers, probably the Old Spanish Pointer which was brought to England in 1713. Crossings with the Foxhound and Greyhounds (see pp. 232–3) increased their pace, with the Bloodhound (see pp. 196–7) also contributing to the mix, enhancing the Pointer's scenting ability.

PERSONALITY

Friendly, alert and sensitive, the Pointer is easy to train and enjoys working closely with people. It possesses considerable stamina and agility.

HEALTH AND CARE

Pointers can suffer from a hereditary condition called neurotropic osteopathy, which becomes evident between 3–9 months of age. Affected individuals start gnawing at their toes and seem insensitive to the resulting pain, although a much more common cause of toe-biting is tiny harvest mites.

AS AN OWNER

Pointers are easy to train. Puppies may even start to display their characteristic pointing stance as young as 8 weeks, with a front foot raised, the head extended and the tail quivering with concentration.

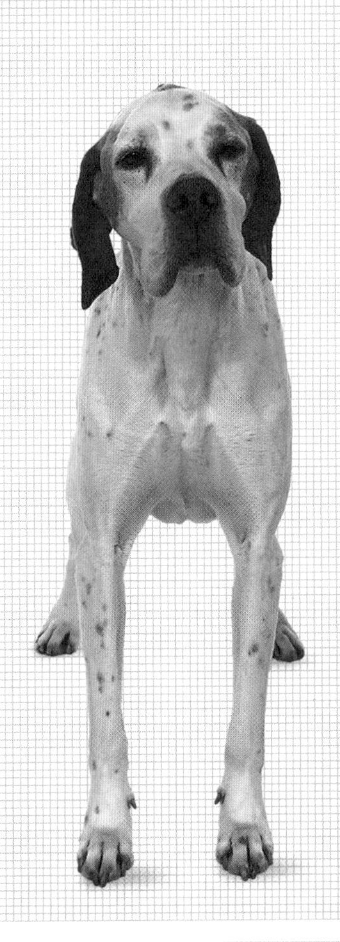

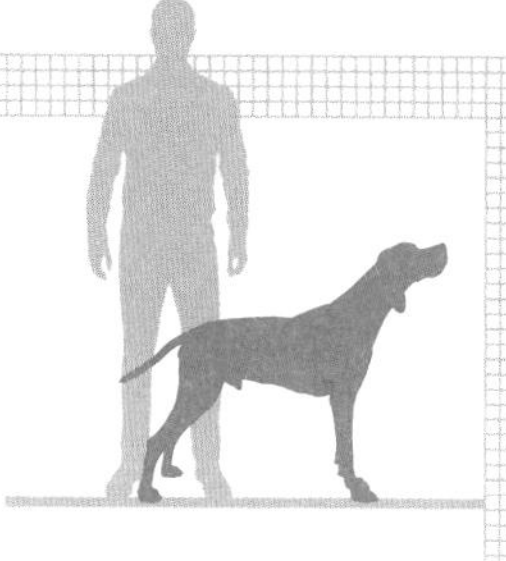

Specification

This dog has an athletic build and a short, sleek coat. The muzzle is broad with wide nasal passages, assisting its tracking skills. The upper lip hangs down below the level of the lower jaw. Thick pads protect the feet, cushioning their movement over rough ground.

RECOGNITION S
North America, Britain and FCI member countries

LIFESPAN
11–13 years

COLOUR
Solid lemon, orange, liver or black, or any combination with white

HEAD
Medium-width with a deep muzzle

EYES
Good size, round and dark

EARS
Set level with the eyes, hanging down to below the lower jaw

CHEST
Deep but not wide, with a bold breastbone

TAIL
Tapered along its length, with no curl; can reach down to the hocks, but not carried between the legs

WEIGHT
Dogs 25–34kg (55–75 lb); bitches 20.5–29.5kg (45–65 lb)

CHILD FRIENDLINESS

GROOMING

FEEDING

EXERCISE

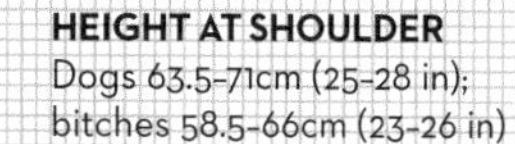
HEIGHT AT SHOULDER
Dogs 63.5–71cm (25–28 in); bitches 58.5–66cm (23–26 in)

GIANT SCHNAUZER

The Giant Schnauzer is every bit as solid and dependable as its physique suggests. Alert by nature, it is a good guardian, with its sheer size likely to deter intruders. This breed was developed in Germany from the Schnauzer with Great Danes and Rottweilers (see pp. 236–7 and 190–1) also both used in the breeding mix. Giant Schnauzers originally worked as cattle drovers in the area around Munich in southern Germany.

PERSONALITY

Intelligent, adaptable, and easy to train, the Giant Schnauzer makes a reliable, good-tempered companion. Spirited and powerful, these dogs display good stamina and will be keen to exercise outdoors irrespective of the weather.

HEALTH AND CARE

This breed has a high incidence of hip dysplasia. This is an inherited weakness of the back legs in which the cup-shaped part of the hip joint is too shallow to allow the rounded head of the femur to fit properly into the socket. Screening of breeding stock is helping to reduce its incidence.

AS AN OWNER

This breed does not shed conventionally, and unwanted hair will need stripping out roughly every six months. This job is best done by a professional dog groomer. Be careful with a Giant Schnauzer around cattle in case its dormant herding instincts re-emerge.

Specification

This is a powerfully built dog, with longer hair on its head and a characteristic beard. The 'salt and pepper' colour combination describes hairs banded black and white along their length, ideally creating a medium grey shade, although this can vary from dark iron-grey to a silvery colour.

RECOGNITION W
North America, Britain and FCI member countries

LIFESPAN
10-12 years

COLOUR
Black or salt and pepper

HEAD
Rectangular and elongated, with a strong muzzle

EYES
Oval, dark brown and medium-sized

EARS
Set high, V-shaped and medium-length; carried close to the head

CHEST
Medium-width

TAIL
Set high; carried high when the dog is alert

WEIGHT
32-35kg (70-77 lb)

GROOMING

FEEDING

EXERCISE

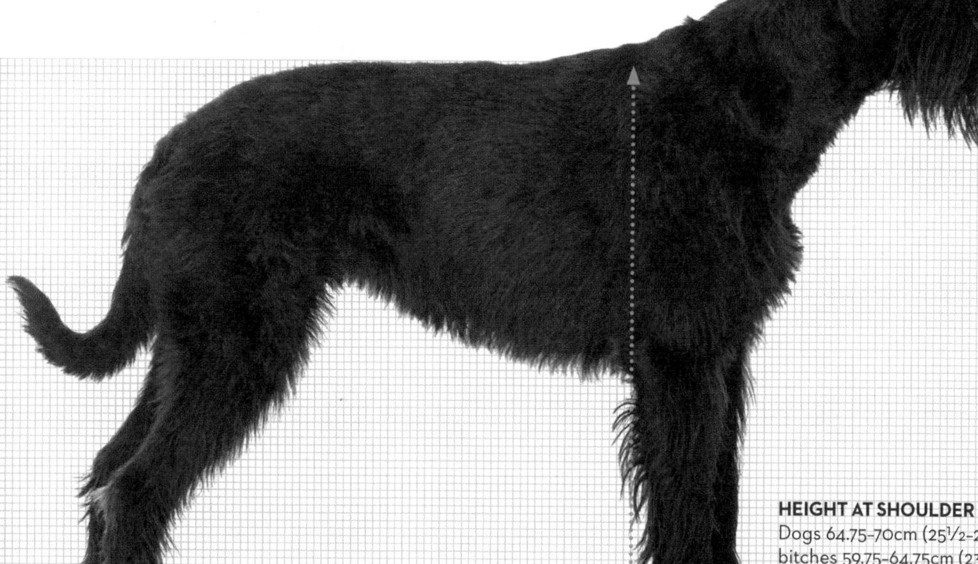

HEIGHT AT SHOULDER
Dogs 64.75-70cm (25½-27½ in); bitches 59.75-64.75cm (23½-25½ in)

BOUVIER DES FLANDRES

The word 'bouvier' means 'bovine herder', and refers to the Bouvier's original function as a cattle dog in its native Belgium. The imposing Bouvier des Flandres is the last surviving member of a group of Belgian herding dogs, with the other three breeds of this type having become extinct as a result of World War I. The Bouvier des Flandres' ancestry is believed to have involved a range of different breeds, including the Beauceron, while its tousled appearance also suggests some Schnauzer input.

PERSONALITY

Fearless and trustworthy, the breed served as a messenger and ambulance dog in World War I. Bold by nature, the Bouvier des Flandres makes an excellent guard dog and its responsiveness and intelligence has seen it selected as a guide dog.

HEALTH AND CARE

The coat needs little care, nor does the Bouvier des Flandres suffer from any particular hereditary weaknesses. Nevertheless, it is not unusual for puppies to be born without dew claws.

AS AN OWNER

The Bouvier des Flandres is a physically strong and determined breed, but is comparatively easy to train, and usually good-natured.

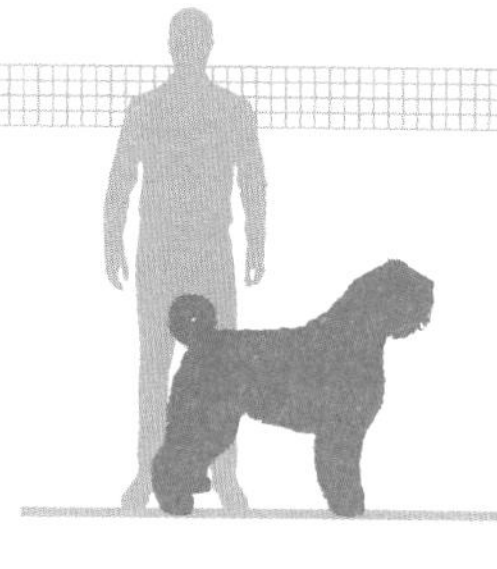

Specification

Powerfully built, with a distinctive beard of longer hair hanging down on each side of its muzzle, the Bouvier des Flandres has a rough-textured, weather-resistant double coat, which is shorter on the skull and the upper back. The undercoat, which provides good insulation, becomes noticeably thicker in winter.

RECOGNITION He
North America, Britain and FCI member countries

LIFESPAN
11–13 years

COLOUR
Varies from fawn to black, although white, chocolate and parti-coloured are not recognized for show purposes

HEAD
Large head with skull longer than the broad, slightly tapered muzzle

EYES
Oval and dark brown

EARS
Set high; creating an alert impression

CHEST
Broad, with the brisket reaching the elbow

TAIL
Set high and carried upright; some puppies are born tail-less

WEIGHT
27–40kg (60–88 lb)

CHILD FRIENDLINESS

GROOMING

FEEDING

EXERCISE

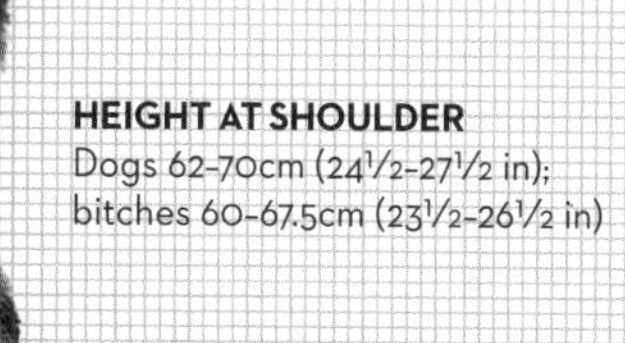

HEIGHT AT SHOULDER
Dogs 62–70cm (24½–27½ in); bitches 60–67.5cm (23½–26½ in)

RHODESIAN RIDGEBACK

Virtually all dog breeds originate from the northern hemisphere, but the Rhodesian Ridgeback was created in Zimbabwe, southern Africa. It is characterized by a very distinctive ridge of hair running down its back, inherited from a native African tribal breed called the Hottentot Dog. Although it is now extinct, this unique characteristic lives on in the Rhodesian Ridgeback, which was bred by early European settlers to hunt lions, using European breeds as well.

PERSONALITY

Bold and fearless, the Rhodesian Ridgeback displays great stamina and requires plenty of exercise.

HEALTH AND CARE

A rare genetic weakness in Ridgebacks can cause the development of a cyst in front of or behind the ridge. This is the result of a developmental abnormality, which is due to the failure of the skin to separate properly from the spinal cord during the puppy's development in the womb, creating a so-called dermoid sinus. Coat care is normally straightforward, with simple brushing being adequate.

AS AN OWNER

Be prepared to take on a strong-willed dog that requires firm training from puppyhood to prevent behavioural problems later. Rhodesian Ridgebacks are fiercely protective of the family and property, and this characteristic needs to be combined with a reliable responsiveness. They may also disagree with other dogs unless well-trained.

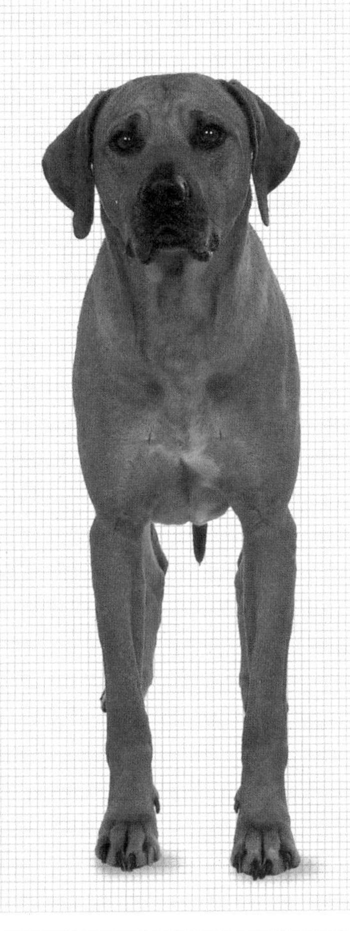

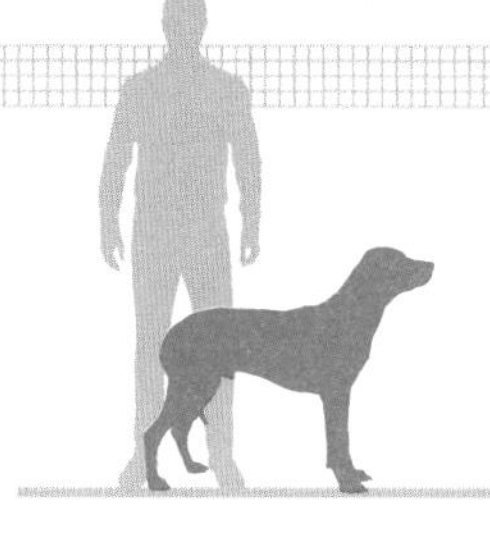

Specification

Ridgebacks are a variable wheaten-tan colour, with small areas of white on the chest and toes being permitted. The coat itself is very sleek and short. The ridge starts behind the shoulders, extending down to the hip bones. The whorls of fur, known as crowns, should be positioned directly opposite each other.

RECOGNITION H
North America, Britain and FCI member countries

LIFESPAN
10–12 years

COLOUR
Wheaten-tan

HEAD
Flat skull and broad between the ears

EYES
Round with an intelligent expression

EARS
Medium-sized and set high

CHEST
Not wide but very deep

TAIL
Strong with a slight upward curve

WEIGHT
38.5kg Dogs (85 lb); bitches 32kg (70 lb)

CHILD FRIENDLINESS

GROOMING

FEEDING

EXERCISE

HEIGHT AT SHOULDER
Dogs 63.5–68.5cm (25–27 in); bitches 61–66cm (24–26 in)

BULLMASTIFF

Massive and immensely powerful, the Bullmastiff is not suited to urban surroundings. The breed was developed on the large country estates of England in the 1800s, at a time when gamekeepers were often being attacked and killed by poachers. It was created by crossings between the Mastiff (see pp. 234–5) and the formidable old-style English Bulldog, as a way of assisting them in order to defend themselves.

PERSONALITY

Today's Bullmastiffs retain a fearless streak, and they are loyal to those in their immediate circle. Equally, they are much more kindly disposed to strangers today, especially those they meet outside their home territory.

HEALTH AND CARE

Possibly because they were bred when the Bulldog was more robust, Bullmastiffs are largely free of inherited weaknesses. They can have an extra incisor tooth in their jaws at the front of the mouth, and may suffer from entropion, a condition affecting the eyelids.

AS AN OWNER

Remain aware and respectful of the immense power of these dogs. They must be trained to walk properly on the leash, so they cannot pull you over, and they are certainly too strong for children to control adequately. This is a breed that will benefit from play while young. Games will help it to improve its coordination – like many larger breeds, Bullmastiff puppies are rather clumsy.

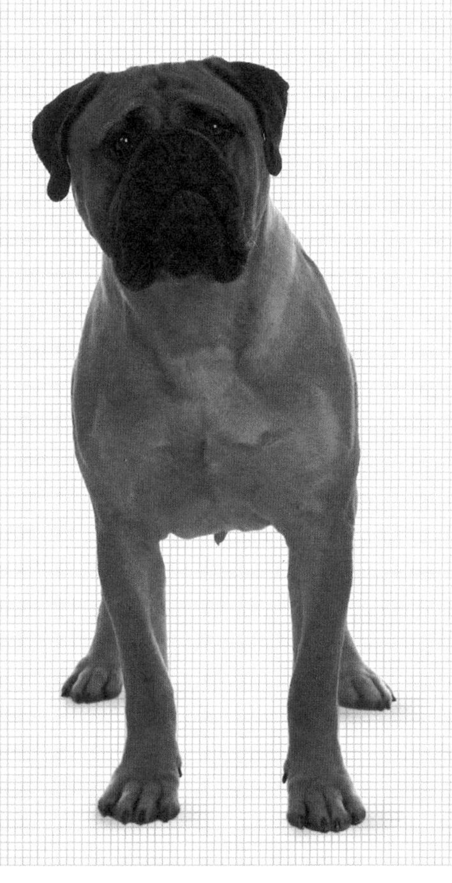

Specification

Lighter fawn colouring is now popular, although the darker brindle shades were originally favoured because these provided better night-time camouflage. The coat is short and feels hard.

RECOGNITION W
North America, Britain and FCI member countries

LIFESPAN
9–11 years

COLOUR
Fawn, red or brindle; some white on the chest

HEAD
Large and broad with a broad, deep muzzle

EYES
Medium-sized and dark

EARS
Set high, widely spaced, and V-shaped, lying close to the cheeks

CHEST
Deep and wide

TAIL
Broad at the base, tapered along its length and set high, reaching the hocks

WEIGHT
Dogs 50–59kg (110–130 lb); bitches 45.5–54.5kg (100–120 lb)

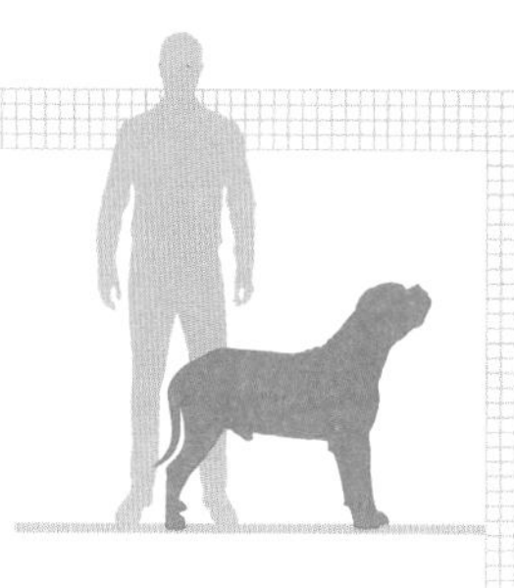

CHILD FRIENDLINESS

GROOMING

FEEDING

EXERCISE

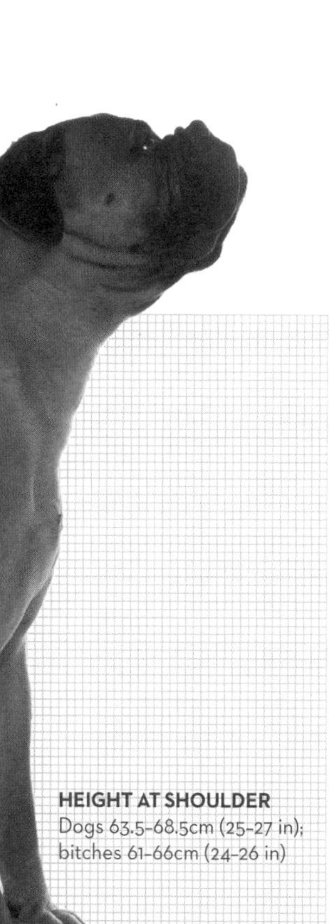

HEIGHT AT SHOULDER
Dogs 63.5–68.5cm (25–27 in); bitches 61–66cm (24–26 in)

AFGHAN HOUND

Named after its home area, centred on modern-day Afghanistan, this sure-footed sight hound has been bred there for centuries, hunting game ranging from hares to antelope and deer. With little natural cover in this barren landscape, pace and stamina were both essential attributes for these hounds. The Afghan was first brought to Britain in the late 1800s, but it was not until the 1920s that the breed started to become more widely known.

PERSONALITY

Rather aloof by nature, the Afghan is a breed that can still retain keen hunting instincts. These hounds are athletic by nature, possessing an independent spirit, and tend to form strong relationships with their owners.

HEALTH AND CARE

Daily grooming is essential to maintain the thick, silky coat, preventing any mats. The breed is prone to various eye problems, including cataracts. Afghan Hounds can also suffer from transient 'blue eye' following vaccination against infectious canine hepatitis.

AS AN OWNER

An Afghan Hound must be carefully trained, to prevent it from chasing small dogs when off the leash. A muzzle may also be recommended as a precautionary measure.

Specification

The early Afghan Hounds varied noticeably from one another in appearance. Those that founded the Bell-Murray strain in the 1920s were from desert areas, being relatively pale in colour and small. The stockier, darker Ghazni strain consisted of individuals with longer coats that originated from more mountainous areas, although no such divisions exist in modern bloodlines.

RECOGNITION H
North America, Britain and FCI member countries

LIFESPAN
10–12 years

COLOUR
No restrictions on colour; white patches not favoured

HEAD
Refined, with good length and a silky topknot

EYES
Dark and almond-shaped, being almost triangular

EARS
Long, and hang level with outer corners of the eyes

CHEST
Deep and narrow

TAIL
Not set very high and curved at the tip

WEIGHT
Dogs approximately 27kg (60 lb); bitches about 22.5kg (50 lb)

CHILD FRIENDLINESS

GROOMING

FEEDING

EXERCISE

HEIGHT AT SHOULDER
Dogs 66–71cm (26–28 in); bitches 61–66cm (24–26 in)

DOBERMANN

The Dobermann's elegant looks and strong character make it a good choice for someone who is particularly interested in dog-training. This breed is less suitable as a family pet, especially alongside young children, because it can sometimes be possessive of toys and short-tempered. Dobermanns will form a close bond on a one-to-one basis, and have been used successfully as police and military dogs because of the relationships they form with their handlers. This breed was created in the late 1800s by a German tax collector, Louis Dobermann, for protection. German Pinschers played a part in its development.

PERSONALITY

Bold, determined, fearless and loyal to its owner, the Dobermann has become much friendlier since its early working days, when it was used to intimidate reluctant payers and would-be thieves.

HEALTH AND CARE

The Dobermann is vulnerable to skin problems, which often result in inflammation and irritation. These may develop as a reaction to allergies or intolerances. An allergic reaction to fleabites is not unusual, so treat your dog regularly against these parasites.

AS AN OWNER

Buy a Dobermann puppy only from a reputable breeder, and train it carefully. Be wary of taking on an adult Dobermann that needs rehoming for any reason – its temperament may be a problem, even if it initially seems friendly.

Specification

The Dobermann has a sleek, tight-fitting coat, with rust-coloured markings above each eye. There are also patches of rust on the muzzle and upper chest, extending to the legs and feet and below the tail.

RECOGNITION W
North America, Britain and FCI member countries

LIFESPAN
10–12 years

COLOUR
Black, red, blue or fawn (Isabella), plus rust markings

HEAD
Long, resembling a blunt wedge in profile

EYES
Almond-shaped, with colouration corresponding to the coat

EARS
Small and set high

CHEST
Broad, with an evident fore chest

TAIL
Forms a continuation of the back, carried slightly above the horizontal

WEIGHT
30–40kg (66–88 lb)

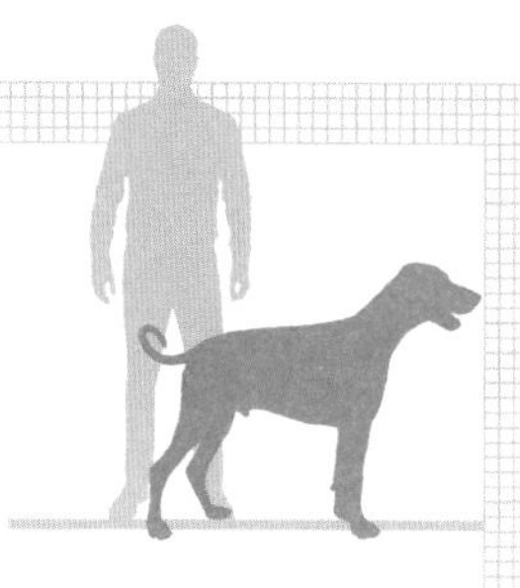

CHILD FRIENDLINESS

GROOMING

FEEDING

EXERCISE

HEIGHT AT SHOULDER
Dogs 66–71cm (26–28 in); bitches 61–66cm (24–26 in)

AKITA INU

These large, powerful dogs need plenty of space, and this is certainly not a breed to contemplate unless you have a spacious back garden plus the time and energy to take your dog for regular strenuous walks. This breed was originally developed in the 1600s for hunting bears in the Japanese province after which it is named, on the island of Honshu. It attracted international attention thanks to a dog named Hachiko, which always accompanied his owner to and from the train station each day. Sadly, the man died at work, and for the rest of Hachiko's life, the dog went to meet the train in the vain hope that he would return.

PERSONALITY

This is an immensely loyal breed, with a strong, protective nature – these dogs were even left to guard young children in their mother's absence. Intelligent and versatile, they are brave and not easily intimidated.

HEALTH AND CARE

Their great strength means Akita Inus must be well trained from puppyhood. Although they need more grooming in the spring, when shedding their winter coat, grooming care is straightforward. Be cautious of Akita Inus when meeting other dogs, especially dominant breeds – if challenged, they are unlikely to back down.

AS AN OWNER

Large and imposing in build, this breed requires sound training from an early age, as well as good socialization. Akita Inus are rather strong-minded and not naturally well-disposed to either strangers or other dogs.

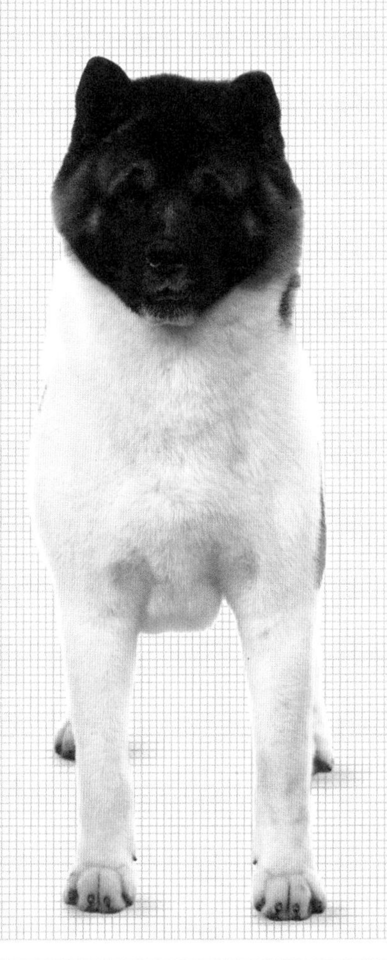

Specification

The Akita Inu is large and strongly built, with pricked ears that give it an alert expression. The neck is thick and muscular, and the shoulders are powerful. The dog is protected against the cold by a double-layered coat, which includes a thicker, soft undercoat. The outer layer is harsh and is longest over the withers and rump.

RECOGNITION W
North America, Britain and FCI member countries

LIFESPAN
10–12 years

COLOUR
No restrictions

HEAD
Large, with square, powerful jaws

EYES
Small and dark brown

EARS
Small, triangular and rounded at the tips; set high

CHEST
Deep and wide

TAIL
Large, set high and carried over the back or to the side in a curl

WEIGHT
34–50kg (75–110 lb)

CHILD FRIENDLINESS

GROOMING

FEEDING

EXERCISE

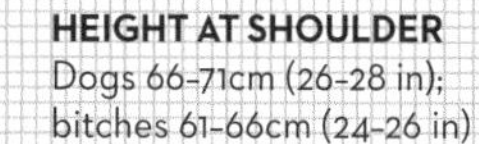

HEIGHT AT SHOULDER
Dogs 66–71cm (26–28 in); bitches 61–66cm (24–26 in)

NEAPOLITAN MASTIFF

This huge, lumbering breed has been likened to a canine hippopotamus! The Neapolitan Mastiff is not a dog for the faint-hearted – it is immensely strong, with its sheer size meaning that it should only be kept in spacious surroundings, and it has an appetite to match its physique. This is one of the oldest surviving breeds in the world, dating back some 2,300 years ago to the time of Alexander the Great, with its ancestors said to originate from northern India.

PERSONALITY

Today's Neapolitan Mastiff is far less aggressive than its predecessors, but it will be protective of its home surroundings. Despite their fearsome appearance, these dogs are usually calm, friendly and placid, and respond well to training.

HEALTH AND CARE

The facial skin folds can become infected, particularly if food deposits accumulate here, so be sure to keep this area clean.

AS AN OWNER

Neapolitan Mastiffs drool heavily, especially around food or in hot weather. Be prepared to carry a cloth or paper towels around to wipe your dog's face and prevent it from depositing pools of saliva on carpets or soft furnishings. These massive dogs are susceptible to the heat stroke.

Specification

A massive, wrinkled, rounded head, with a dewlap extending from the lower jaw down to the middle of the neck, is striking. The neck is powerful and the body is strong and muscular. The fur is dense, glossy and feels quite rough.

RECOGNITION W
North America, Britain and FCI member countries

LIFESPAN
10–12 years

COLOUR
Tawny, mahogany, grey and black; tan brindling, and some white markings allowed

HEAD
Large, with a muzzle whose breadth matches its length

EYES
Set deep, with drooping lids

EARS
Medium-sized and triangular; held against the cheeks

CHEST
Powerful, broad and deep

TAIL
Broad at its base and tapered along its length; can be raised just above the horizontal

WEIGHT
Dogs 68kg (150 lb); bitches 50kg (110 lb)

CHILD FRIENDLINESS
GROOMING
FEEDING
EXERCISE

HEIGHT AT SHOULDER
Dogs 66–79cm (26–31 in); bitches 61–73.5cm (24–29 in)

IRISH SETTER

The stunning colouring of this setter has guaranteed its popularity, and it has been nicknamed the 'Red' Setter. Unfortunately, although it has a kindly disposition, its background and behaviour mean that it is not ideal for the urban environment. A native of Ireland, this Setter was developed from the Old Spanish Pointer crossed with various types of spaniels. It was originally used to hunt birds, ranging from pheasants to wild ducks, driving them into nets or flushing them for an accompanying falcon to catch overhead. The breed was later crossed with Borzois (see pp. 228–9) to improve its elegance and speed, becoming taller as well.

PERSONALITY

A true extrovert, the Irish Setter is an enthusiastic, energetic breed. These setters can be wayward, however, especially while they are young and not fully trained, often displaying a tendency to run off.

HEALTH AND CARE

Puppies may be affected by an obstruction in the gullet, caused by the incorrect development of the aorta, the body's major artery. It makes swallowing difficult, causing an affected puppy to regurgitate food. Surgery may be helpful.

AS AN OWNER

Do not be seduced by this dog's beauty. Irish Setters need plenty of daily exercise – they can otherwise become destructive. Be patient when training your Irish Setter; this breed learns slowly.

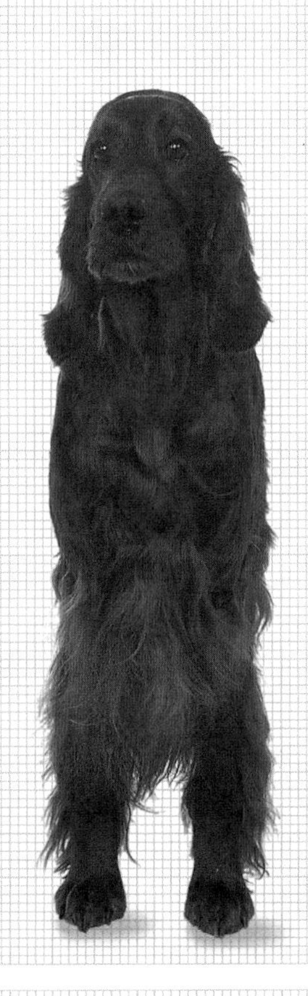

Specification

Aside from their colouring, their smooth, powerful and elegant gait is impressive. The coat is at its shortest on the front of the forelegs and the head, and longer – and flat – elsewhere on the body.

RECOGNITIONS
North America, Britain and FCI member countries

LIFESPAN
11–13 years

COLOUR
Mahogany to rich shades of chestnut red

HEAD
About twice as long as the width between the ears

EYES
Well-spaced, medium-sized and almond-shaped

EARS
Set at or below the level of the eye, hanging close to the head

CHEST
Deep, extending to the elbows

TAIL
Broad at the base and tapered to a point

WEIGHT
Dogs 32kg (70 lb); bitches 27kg (60 lb)

CHILD FRIENDLINESS
GROOMING
FEEDING
EXERCISE

HEIGHT AT SHOULDER
Dogs 68.5cm (27 in); bitches 63.5cm (25 in)

PYRENEAN MOUNTAIN DOG

Also known as the Great Pyrenees, this ancient mastiff breed has an impressive appearance, and makes an amiable companion. The breed was accorded the title of the Royal Dog of France by Louis XIV in the 17th century, and it has attracted many devotees since then, partly as the result of numerous recent film appearances. These huge dogs were originally used to guard flocks in the Pyrenean mountains between France and Spain.

PERSONALITY

Affectionate and devoted to its family, the Pyrenean remains watchful and does not accept visitors readily. The breed has a somewhat independent streak, proving both intelligent and resourceful.

HEALTH AND CARE

The double dew claws should be trimmed back regularly. If they are not removed, they will curl around into the fleshy pad, if they are not removed.

AS AN OWNER

These dogs need plenty of indoor space. Provide a clear floor area, and remove objects that could be swept off low surfaces by your dog's tail. Invest in a large-size beanbag, for your dog to stretch out on and sleep. Conventional dog beds are often not very comfortable for large dogs in general.

Specification

Massive and unmistakable, the Pyrenean Mountain Dog is well protected by a double coat – the outer layer is flat and thick, while the undercoat is fine and dense. Unusually the Pyrenean has double dew claws on its hind legs, although they have no practical benefit.

RECOGNITION W
North America, Britain and FCI member countries

LIFESPAN
10–12 years

COLOUR
Solid white, or white with markings in shades of grey, badger, reddish brown or tan

HEAD
Wedge-shaped, with the muzzle and skull of similar size

EYES
Almond-shaped, rich dark brown and medium-sized

EARS
V-shaped, small to medium with rounded tips; usually carried low and flat

CHEST
Broad

TAIL
Well-plumed; can extend down to the hocks; may be carried above the back

WEIGHT
Dogs 45.5kg (100 lb); bitches 38.5kg (85 lb)

CHILD FRIENDLINESS

GROOMING

FEEDING

EXERCISE

HEIGHT AT SHOULDER
Dogs 68.5–81cm (27–32 in); bitches 63.5–73.5cm (25–29 in)

ST BERNARD

The St Bernard is named after Bernard de Mentho, founder of the Bernardine Hospice in the Swiss Alps, where the breed was developed. It is well known for rescuing people stranded in blizzards, and proves to be a trustworthy and affectionate companion. St Bernards get along well with children, but they are not particularly effective as guard dogs. They are descended from Alpine mastiffs, bred in the region from Roman times.

PERSONALITY

Well-disposed towards people, the St Bernard is determined and possesses a keen sense of smell. These dogs may often show an independent streak, having been bred to take the initiative in searches.

HEALTH AND CARE

This breed suffers from various inherited health problems, including blood-clotting disorders, and issues around the eyes and eyelids. Older dogs are vulnerable to cancerous bone tumours in the legs, known as osteosarcomas. Lameness is an early sign, and rapid diagnosis is essential to prevent the cancer spreading to the lungs.

AS AN OWNER

St Bernards can be clumsy around the house – in particular their broad, swooshing tails are a hazard, sweeping objects off any low surface. This is a breed that needs plenty of space to be comfortable. Like other square-muzzled mastiff breeds, it also has a tendency to drool.

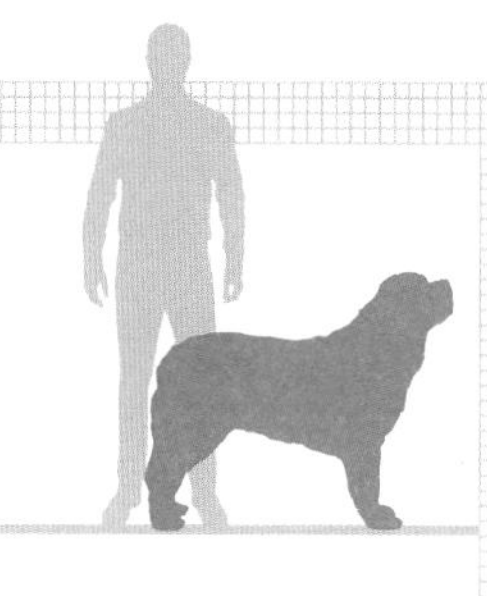

Specification

Longhaired and shorthaired forms of the breed exist – and both possess dense fur that protects well against the cold.

RECOGNITION W
North America, Britain and FCI member countries

LIFESPAN
9–11 years

COLOUR
White and variable red combinations, including brindle

HEAD
Massive and wide with high cheek bones and a short muzzle

EYES
Medium-sized and dark brown

EARS
Set high, held away from the head at their base, but dropped down alongside the head

CHEST
Deep, but not extending down to the elbows

TAIL
Broad and long, hanging straight down at rest; otherwise, carried upward

WEIGHT
50–91kg (110–200 lb)

CHILD FRIENDLINESS

GROOMING

FEEDING

EXERCISE

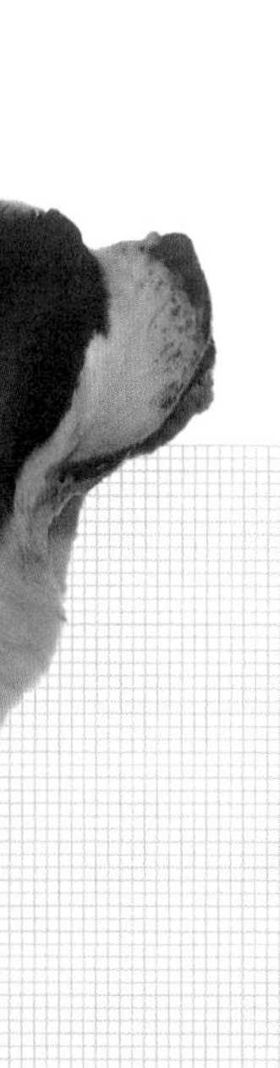

HEIGHT AT SHOULDER
Dogs 70cm (27½ in) or more; bitches 64.75cm (25½ in) or more

NEWFOUNDLAND

Its calm, gentle temperament means this breed makes an ideal companion if you have space available, although like other large dogs, Newfoundlands are expensive to keep, because of their hearty appetites. The Pyrenean Mountain Dog (see pp. 222–3) probably played a part in its development, as may native Inuit dogs in its North American homeland. The Newfoundland used to pull carts, but once its exceptional swimming abilities were recognized, it began to work with fishermen, helping to haul in their nets.

PERSONALITY

A calm and loyal companion, this dog has long been portrayed – with justification – as a protector of children. The Newfoundland's innate intelligence and relaxed nature mean that these dogs adapt well to new situations.

HEALTH AND CARE

Puppies must be checked up to the age of 6 months for subaortic stenosis, which leads to a narrowing of the aorta. Affected dogs become breathless after exercise, and may faint. If untreated, the condition can even cause sudden death.

AS AN OWNER

Any dog, especially a large one, is a potential liability if is not properly trained from puppyhood. The Newfoundland is instinctively keen to please, so training is usually straightforward.

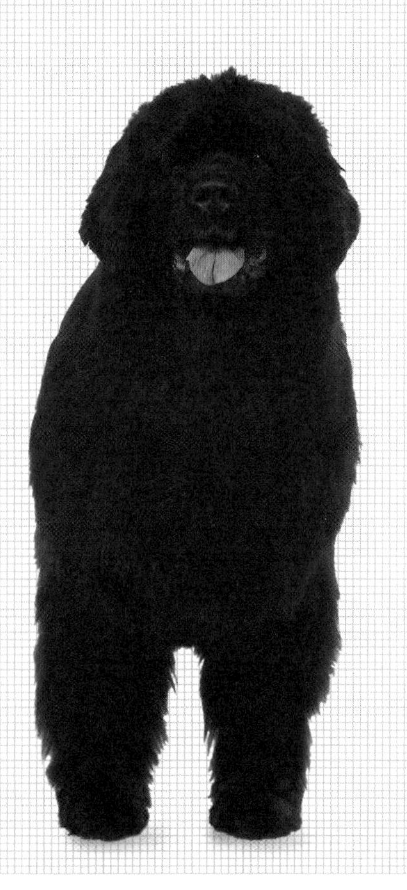

Specification

This is a powerful breed with a heavy, water-resistant double coat, which becomes thinner in the summer. The toes are webbed, aiding its swimming ability. The black-and-white form is called the Landseer, commemorating Queen Victoria's favourite dog artist, Edwin Landseer. Sometimes recognized as a separate breed, it is nevertheless identical in all respects other than its colouring.

RECOGNITION W
North America, Britain and FCI member countries

LIFESPAN
10–12 years

COLOUR
Brown, grey, black or black and white

HEAD
Large, with a broad skull and a broad, deep muzzle

EYES
Dark brown

EARS
Triangular-shaped and small, with rounded tips

CHEST
Deep and full, reaching the elbows

TAIL
Broad and powerful

WEIGHT
Dogs 59–68kg (130–150 lb); bitches 45.5–54.5kg (100–120 lb)

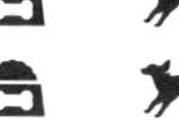

CHILD FRIENDLINESS

GROOMING

FEEDING

EXERCISE

HEIGHT AT SHOULDER
Dogs 71cm (28 in); bitches 66cm (26 in)

BORZOI

For sheer elegance, the Borzoi is a breed with few rivals. Its narrow head reveals that it is a sight hound, and its name derives from the Russian word 'borzyi', meaning 'swift', reflecting its remarkable pace. Borzois are probably descended from Greyhound-type stock crossed with native long-coated sheepdogs. Traditionally these hounds worked in pairs, hunting wolves, and were known as Russian Wolfhounds. They became popular with the Russian Royal Family, and this almost led to their demise after the Russian Revolution of 1917. Luckily by then, Borzois were already represented elsewhere in Europe.

PERSONALITY

Rather aloof towards strangers, the Borzoi has a reputation for bravery, traditionally wrestling wolves to the ground and holding their quarry until a huntsman could inflict a fatal blow.

HEALTH AND CARE

Missing teeth can be an issue, simply failing to develop in the jaws, and are a show fault. This breed is more easily trained than many hounds, reflecting its ancestry. Regular grooming of the longer areas of fur on the coat is necessary.

AS AN OWNER

Only choose this breed if you have adequate space nearby for exercise. Developed to hunt in couples, Borzois generally agree well, but space can be an issue. This is a sensitive breed, and makes a quiet companion, not predisposed to barking loudly even at strangers.

Specification

The Borzoi's coat is long and slightly wavy, with a silky texture, and is often white in colour. Their long hind legs help them to run with considerable grace and power.

RECOGNITION H
North America, Britain and FCI member countries

LIFESPAN
10–12 years

COLOUR
No restrictions

HEAD
Long and narrow, slightly domed above

EYES
Dark in colour, creating an intelligent impression and set obliquely

EARS
Small in size, raised when alert

CHEST
Narrow and deep

TAIL
Long, set low, and carried in a curve below the level of the back

WEIGHT
Dogs 34–47.5kg (75–105 lb); bitches 27–43kg (60–95 lb)

CHILD FRIENDLINESS

GROOMING

FEEDING

EXERCISE

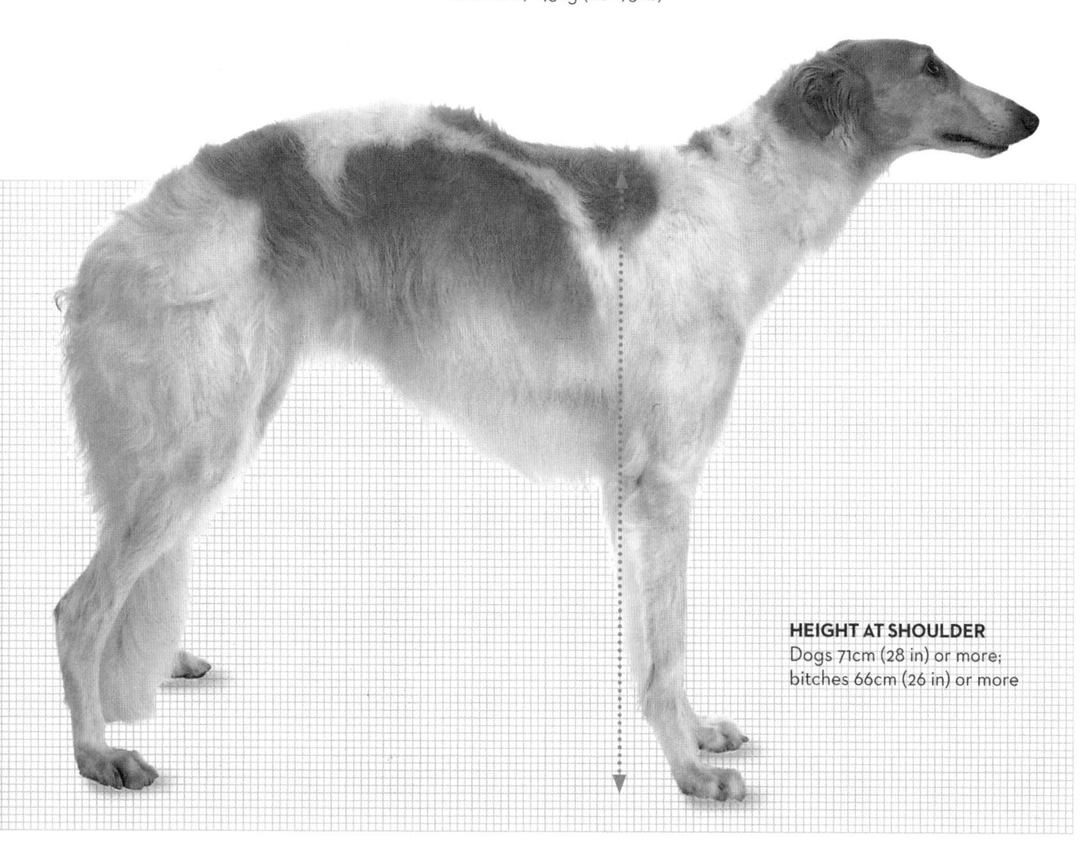

HEIGHT AT SHOULDER
Dogs 71cm (28 in) or more; bitches 66cm (26 in) or more

RUSSIAN BLACK TERRIER

Now regarded as a working dog, reflecting its original use, the Russian Black Terrier proves naturally reserved by nature, and yet is friendly towards people it knows well. As its name suggests, it was first bred in the former Soviet Union, during the 1940s, at the Central School of Cynology Specialists, located just outside Moscow. The breed was developed from, among others, the Giant Schnauzer, the Rottweiler and the Airedale Terrier (see pp. 204–05, 190–91 and 164–5), with great emphasis being placed on its soundness.

PERSONALITY

A determined guard dog with a steadfast nature, the Russian Black Terrier is responsive to training, and, once trained, it is easy to control. Intelligent and with a dependable temperament, it makes a loyal companion.

HEALTH AND CARE

Confident by nature, the Russian Black Terrier is a breed that learns quickly what is expected and thrives on an established routine. As with many larger working breeds, there is a natural variance in size and appearance between male dogs and bitches. The breed needs to be trimmed slightly for show purposes, to enable judges to study the physique more easily, but the overall effect must remain natural.

AS AN OWNER

This is a breed created to work closely with people. Provided that you invest the time, you will have a very loyal companion, but try to involve other family members in the dog's training, so it will respond well to them as well.

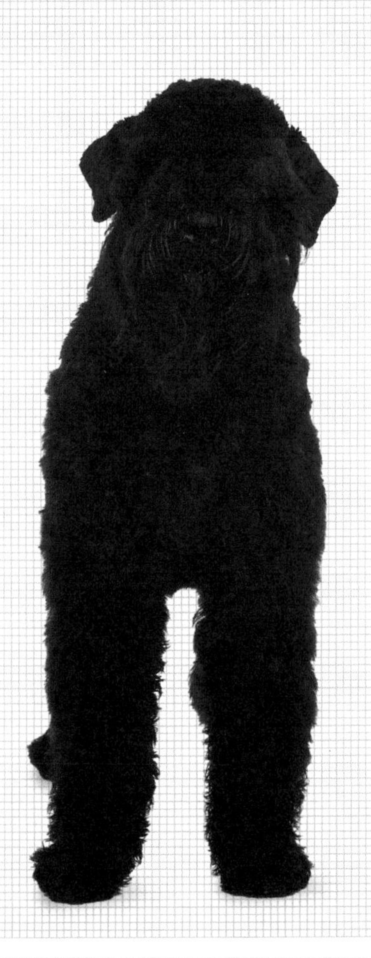

Specification

This is a powerful breed, black in colour and with a tousled, wiry coat. Its overall appearance suggests that the Giant Schnauzer was the dominant influence in its development. The outer coat layer is up to 10cm (4 in) long. The nose and lips are also black, and there is sometimes a black mark on the tongue.

RECOGNITION W
North America, Britain and FCI member countries

LIFESPAN
10–13 years

COLOUR
Solid black

HEAD
Wide head; the muzzle is slightly shorter

EYES
Oval, medium-sized and dark

EARS
Set high, small and triangular

CHEST
Wide and deep

TAIL
Thick, set high and carried above the vertical

WEIGHT
38.5–63.5kg (85–140 lb)

CHILD FRIENDLINESS

GROOMING

FEEDING

EXERCISE

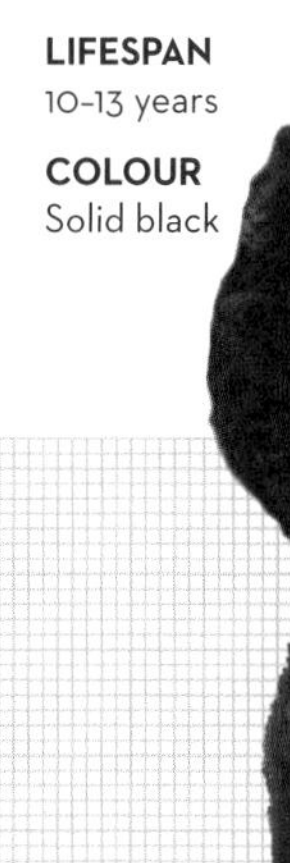

HEIGHT AT SHOULDER
Dogs 68.5–76cm (27–30 in); bitches 66–73.5cm (26–29 in)

GREYHOUND

The Greyhound is the fastest breed of domestic dog, capable of sprinting at 64kph (40 mph) over short distances. Hounds with an almost identical appearance to the contemporary Greyhound can be seen on the tombs and other art of ancient Egypt, dating back more than over 6,000 years. The breed has been highly valued for its hunting abilities, and it is often crossed with sheepdogs to create Lurchers, favoured originally as poachers' companions and, increasingly, as pets. They possess the Greyhound's pace combined with the instinctive responsiveness of a sheepdog.

PERSONALITY

Gentle and friendly by nature, Greyhounds have a quiet temperament. It is usually possible to train them easily, and they can be very affectionate. Couples get along well together.

HEALTH AND CARE

Ex-racing Greyhounds sometimes have residual injuries, which may be behind the decision to retire them from the track. Never exercise these hounds after a meal, in order to avoid bloat, a condition in which the stomach twists, trapping air.

AS AN OWNER

Always muzzle Greyhouns when you let them off for a run – their chasing instinct is very strong – so as to protect other animals from being caught. Greyhounds only require short runs, often looping around in a circle rather like being on a racetrack. The sleek short coat means that grooming is straightforward.

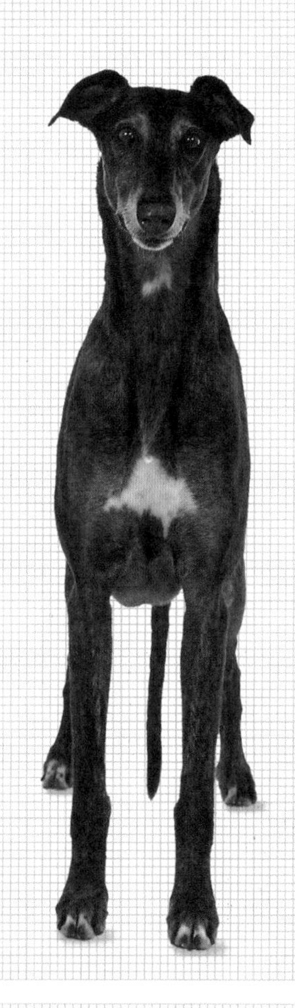

Specification

Greyhounds have a deep chest which gives good lung capacity. This is vital for their athletic performance. Slight differences in appearance are evident between more refined show greyhounds and racing stock.

RECOGNITION H
North America, Britain and FCI member countries

LIFESPAN
10-12 years

COLOUR
No restrictions

HEAD
Long, narrow and relatively wide between the ears

EYES
Bright, intelligent and dark

EARS
Small, set high and fine-textured; usually semi-folded, but raised if excited

CHEST
Deep and relatively wide, with well-sprung ribs

TAIL
Set low, long and tapering, with a slight curve and carried low

WEIGHT
Dogs 29.5-32kg (65-70 lb); bitches 27-29.5kg (60-65 lb)

CHILD FRIENDLINESS

GROOMING

FEEDING

EXERCISE

HEIGHT AT SHOULDER
Dogs 71-76cm (28-30 in); bitches 68.5-71cm (27-28 in)

MASTIFF

Also called the Old English Mastiff, this breed was originally kept as a fighting dog. The ancestors of these dogs were already being bred in Britain more than 2,000 years ago, at the time of the Roman invasion. It is still immensely strong, with proper training from an early age being essential.

PERSONALITY

The Mastiff's temperament has changed considerably through the centuries, and these dogs are much friendlier nowadays. Even so, they are still capable of displaying strong territorial instincts and do not take readily to strangers.

HEALTH AND CARE

Mastiffs often show signs of discomfort in hot weather, panting and drooling heavily. They should be kept cool and should not be exercised during the hottest part of the day. The skin folds on the head sometimes develop localized infections. Due to the dog's great weight, pressure sores in the form of hard, hairless pads may develop on the elbows where the dog lies down, and although often not causing pain, they are unsightly.

AS AN OWNER

Mastiffs have prodigious appetites, but must not be allowed to become overweight. A careful diet and plenty of exercise are essential. Grooming is straightforward.

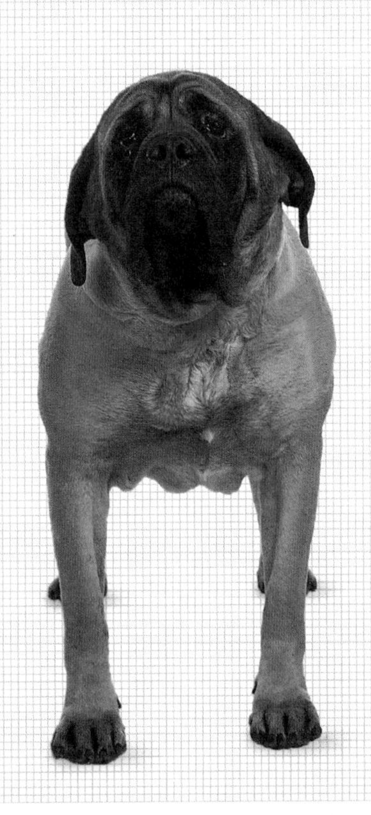

Specification

This breed is significantly bigger and heavier than the Bullmastiff (see pp. 210–11). It has a characteristically dignified appearance, with deep flanks that emphasize its strong physique. Mastiffs have a double coat, dense and short underneath and coarser on top, with the neck being broad and muscular.

RECOGNITION W
North America, Britain and FCI member countries

LIFESPAN
10–12 years

COLOUR
Apricot or fawn, or brindled with either apricot or fawn

HEAD
Broad and flat between the ears, and a short, broad muzzle

EYES
Well-spaced, medium-sized and dark brown

EARS
Small and V-shaped with rounded tips

CHEST
Rounded, wide and deep, reaching the elbow

TAIL
Set high and tapered to the level of the hocks; not carried above the back

WEIGHT
79–86kg (175–190 lb)

CHILD FRIENDLINESS

GROOMING

FEEDING

EXERCISE

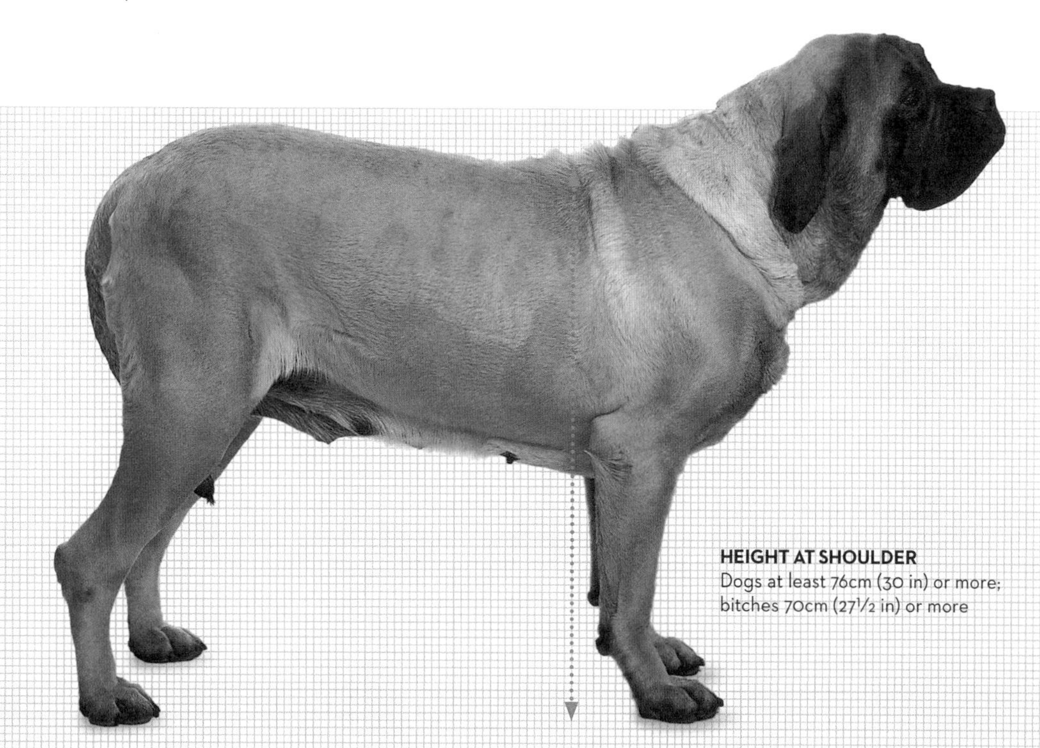

HEIGHT AT SHOULDER
Dogs at least 76cm (30 in) or more; bitches 70cm (27½ in) or more

GREAT DANE

This is one of the tallest breeds in the world, occasionally measuring over 101.5cm (40 in) at the shoulder – literally as large as a small pony. If you have the space both outdoors and in your home, a Great Dane can make a wonderful pet, with a friendly temperament. Its ancestors were highly prized hunting dogs, with a heavier build.

PERSONALITY

Docile and gentle despite their huge size, Great Danes have an enthusiastic, playful approach to life, living well alongside teenagers.

HEALTH AND CARE

Great Danes can suffer from spondylomyelopathy or 'Wobbler Syndrome'. This inherited condition affects the cervical vertebrae in the neck, through which the spinal cord runs. Symptoms can range from a mild weakness of the hind legs to paralysis. Sadly, Great Danes are relatively short-lived, being vulnerable to arthritis and bone cancer as they age.

AS AN OWNER

Great Dane puppies develop slowly and, as with other giant breeds, must not be over exercised, to guard against joint weaknesses later. The breed can also develop bloat, a serious condition resulting from a buildup of gas caused by a twisted stomach, and to avoid this they should not be exercised soon after eating.

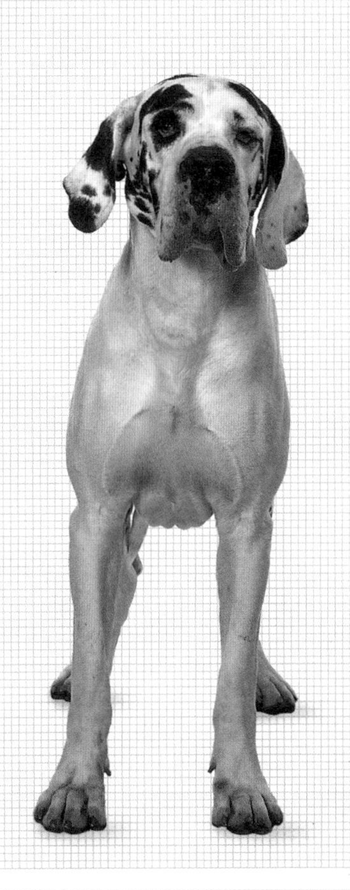

Specification

The breed's colouration includes a striking white form with irregular black patches on the head and body, although not the neck, called the Harlequin. The Mantle can be recognized by a black area resembling a coat, covering the body.

RECOGNITION W
North America, Britain and FCI member countries

LIFESPAN
8–11 years

COLOUR
Fawn, blue, black, Harlequin and Mantle

HEAD
Long and rectangular; the muzzle's length matches that of the skull

EYES
Almond-shaped, medium-sized and dark

EARS
Set high, with the ear folded level with the top of the skull

CHEST
Broad, deep and muscular

TAIL
Broad and tapered along its length to the level of the hocks; not carried above the level of the back

WEIGHT
45.5–54.5kg (100–120 lb)

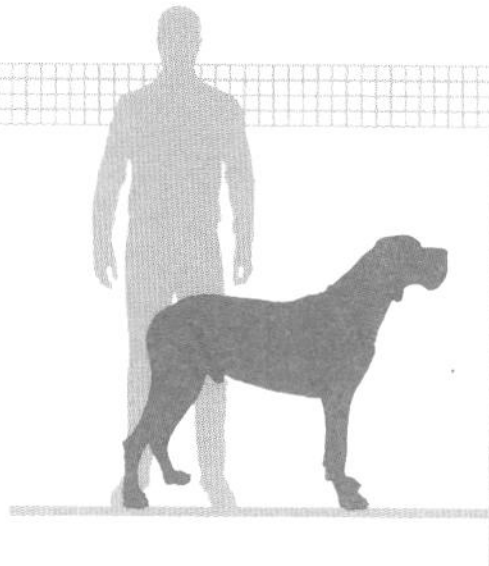

CHILD FRIENDLINESS

GROOMING

FEEDING

EXERCISE

HEIGHT AT SHOULDER
Dogs 76–81cm (30–32 in) or more; bitches 71–76cm) (28–30 in) or more

IRISH WOLFHOUND

The Irish Wolfhound towers over most other breeds, but it is a gentle giant. These huge hounds were originally bred in their Irish homeland to hunt wolves, but when the wolf became extinct there in the 1770s, the future of the hounds became increasingly uncertain. Thanks to crossbreeding with the Deerhound in particular, and other breeds including the Great Dane and the Pyrenean Mountain Dog (see pp. 236–7 and 222–3), the Irish Wolfhound was saved from extinction.

PERSONALITY

This breed is docile and affectionate by nature, but can also be exuberant, particularly when young. These hounds respond well to training, which is vital in view of their considerable size.

HEALTH AND CARE

Avoid overexercising puppies because this can lead to joint problems later in life. Regular, relatively short spells of exercise are much better than more occasional long marathons. The coat needs little care, and it is remarkably damp-proof in wet weather. Older individuals are susceptible to bone tumours, called osteosarcomas. These sometimes necessitate the amputation of the affected limb, but Irish Wolfhounds often adapt surprisingly well.

AS AN OWNER

Plenty of space is essential to keep one of these giants. They can be inadvertently clumsy – their long strong tails can easily sweep things off any low surface. Irish Wolfhounds are best accommodated with access to a secure paddock, where they can run around freely.

Specification

A tall, tousle-haired dog, the Irish Wolfhound has a body profile resembling a Greyhound, allowing it to run very fast. Its long stride gives it an effortless gait, supported by strong, curved nails, which help it to maintain its balance at speed.

RECOGNITION H
North America, Britain and FCI member countries

LIFESPAN
8–10 years

COLOUR
Include white, fawn, red, grey, black and brindle

HEAD
Long and level, with a long muzzle

EYES
Dark

EARS
Small, resembling those of a Greyhound

CHEST
Very deep and moderately broad

TAIL
Long and slightly curved

WEIGHT
Dogs 54.5kg (120 lb); bitches 41kg (90 lb)

CHILD FRIENDLINESS

GROOMING

FEEDING

EXERCISE

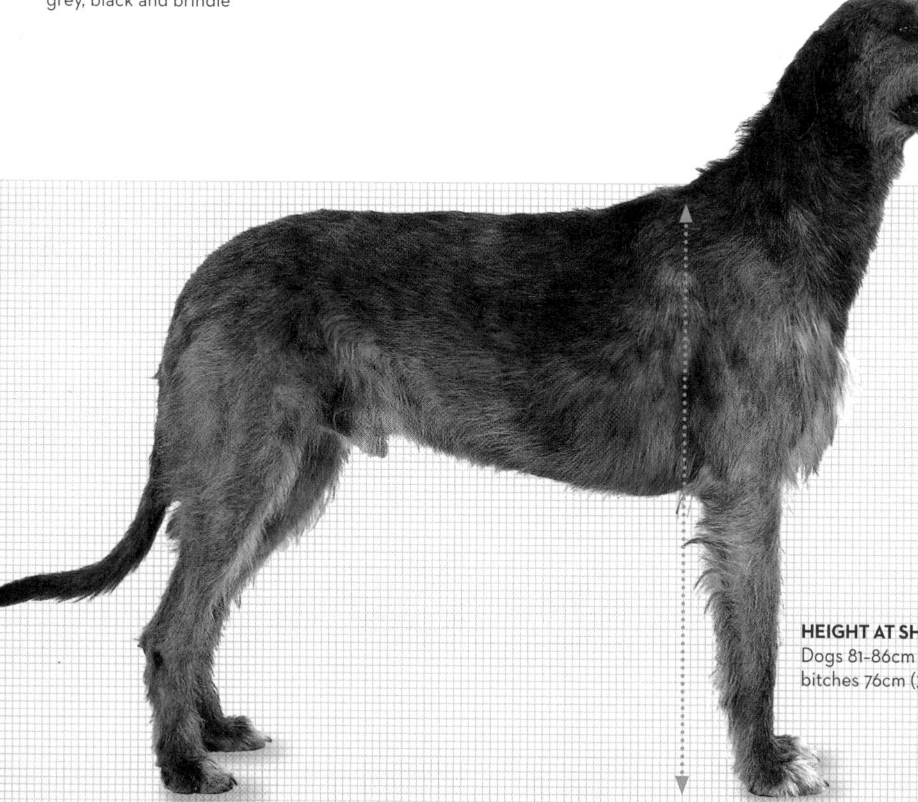

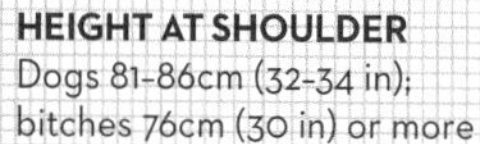

HEIGHT AT SHOULDER
Dogs 81–86cm (32–34 in); bitches 76cm (30 in) or more

COCKAPOO
Cocker Spaniel x Poodle

Not all designer dog crosses have proved to be popular, either because the resulting puppies generally do not look very cute, or because of their personalities. The Cockapoo's popularity, however, is not hard to understand, given that it scores highly on both these points. The size of these dogs depends on the cross itself, with the Miniature Poodle tending to be used most commonly, rather than a Standard (see pp. 114–5) or Toy Poodle. Some have involved English Cocker Spaniels (see pp. 116–7), whereas in other cases, American Cocker (see pp. 100–1) have been used.Records of Cockapoos date back to the 1960s.

PERSONALITY

Inquisitive, enthusiastic and playful, the Cockapoo will prove to be an intelligent, friendly companion.

HEALTH AND CARE

The Cockapoo's coat type varies, even between litter mates, and this affects the grooming requirements of individuals. These dogs can seemingly suffer from separation anxiety quite easily, so train a puppy to be left for periods on its own from an early age. Separation anxiety will cause a dog to bark repeatedly and may also trigger destructive behaviour around the home in your absence.

AS AN OWNER

Aside from their attractive appearance, cockapoos may not shed as much as other dogs, because of their poodle ancestry. They are also relatively easy to train and responsive, being happy to play and chase after balls in a well-fenced garden.

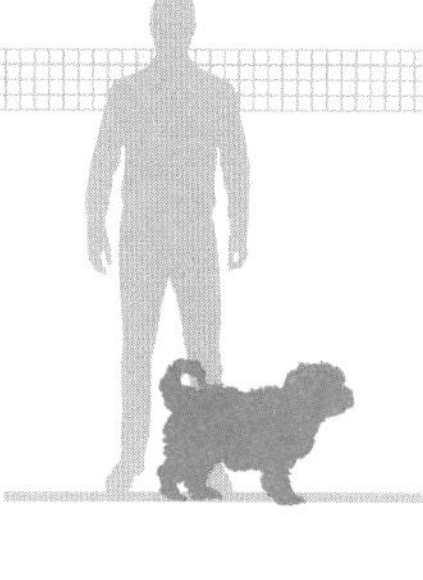

Specification

Variable, as with all designer dogs, although often tends to have the curled coat associated with poodles, rather than the relatively flat coat of cocker spaniels. Apricot and chocolate examples often tend to be favoured. Breeding of cockapoos over several generations has tended to standardize the type of these dogs – starting to turn them into a breed with recognizable characteristics.

RECOGNITION D
Not generally recognized - only by Cockapoo clubs

LIFESPAN
10–12 years

COLOUR
Exists in solid, parti-colours with white areas too

HEAD
Relatively broad skull and corresponding muzzle with prominent nose

EYES
Quite deep set and dark

EARS
Long, wide and set well-back on the skull

CHEST
Deep and medium-width

TAIL
Set high, thick at its base, and forward over the back

WEIGHT
Variable - significantly influenced by size of poodle used in the cross

GROOMING

FEEDING

EXERCISE

HEIGHT AT SHOULDER
Variable - significantly influenced by size of poodle used in the cross

LABRADOODLE
Labrador Retriever x Poodle

The Labradoodle was the first designer dog to be created as the result of a specific breeding programme. Back in 1989, the first crosses of this type were undertaken in Australia, with a view to creating a dog that was intelligent and could be trained to act as a guide dog for the visually impaired, and yet which - most significantly - did not evoke an allergic reaction in susceptible people, thanks to the Poodle input into the bloodline.

PERSONALITY

Playful and intelligent, the Labradoodle makes a good family pet, but beware that it has a natural affinity with water, and likes getting wet if the opportunity presents itself. These dogs tend to have a very sound temperament, and are very responsive to training.

HEALTH AND CARE

There is a thought that designer dogs are healthier than their pure-bred ancestors. Unfortunately though, this is not the case. Labradoodles can be affected with conditions seen in their parent breeds, such as hip dysplasia, eye diseases and others. Testing of breeding stock also tends to be less prevalent in the case of designer dog crossings, which increases the risk of such problems arising in puppies.

AS AN OWNER

There is no guarantee that Labradoodles will actually prove to be hypoallergenic, and the amount of hair that individuals shed is variable, although the chances are they may evoke less of an allergenic reaction, certainly compared with other many other breeds.

Specification

This is variable, with the Labradoodle's size again being influenced by that of the Poodle breed that contributed its development. The coats of some Labradoodles are also much curlier than others.

RECOGNITION D
Not a recognized breed at present

LIFESPAN
10–12 years

COLOUR
Exists in solid, parti-colours with white areas too

HEAD
Relatively broad skull and corresponding muzzle with prominent nose

EYES
Quite deep set, oval and dark

EARS
Long, wide and set high, close to the sides of the face

CHEST
Relatively deep and wide

TAIL
Relatively inconspicuous

WEIGHT
Variable – depends on the breed of poodle used in the cross

HEIGHT AT SHOULDER
Variable – significantly influenced by poodle used in the cross

PUGGLE
Pug x Beagle

Unlike many designer dogs, the Puggle is not descended from Poodle stock, but is the result of crossings between Beagles (see pp. 80–1) and Pugs (see pp. 66–7). They first started to become popular in the 1990s, with their small size, cute appearance and good disposition making them sought-after family pets.

PERSONALITY

Possessing a friendly, playful disposition, the Puggle can also be rather excitable, and may sometimes be difficult to train, having a rather stubborn side to its nature. Puggles form a close bond with those they know well, revealing why they have become one of the most popular designer dogs.

HEALTH AND CARE

Breeding stock should have been screened for hip dysplasia and similar genetic ailments, to which these breeds are susceptible. Puggles are generally not as pre-disposed to eye injuries as Pugs, because their eyes are not so prominent. Both parent breeds are quite gluttonous so take care not to overfeed a Puggle, and weigh it regularly to ensure that it is not becoming obese.

AS AN OWNER

The short coat makes grooming very straightforward. Puggles can often be quite noisy dogs, given to barking and howling regularly, and may suffer from separation anxiety, so they must be properly trained to be left for periods on their own.

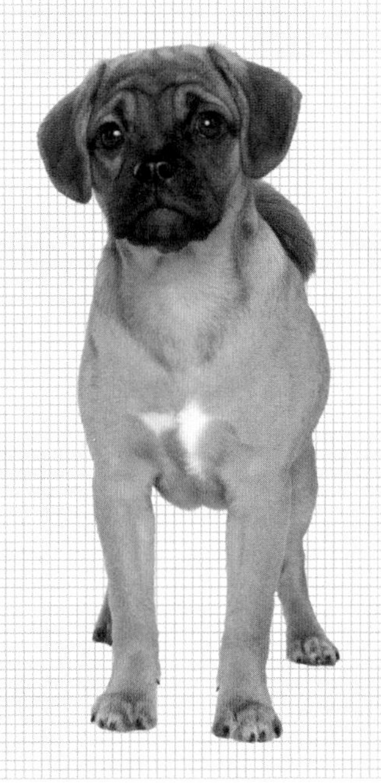

Specification

Being bred from a cross between a Beagle and a Pug, the Puggle is a relative small dog. The muzzle of these crossbreds is much less flattened than that of the Pug, and their colour range is quite variable, with some individuals having black masks on their faces, reminiscent of pugs whereas others display the particolours of their Beagle parent.

RECOGNITION D
Not a recognized breed at present

LIFESPAN
12–14 years

COLOUR
Exists in solid, parti-colours with white areas too

HEAD
Quite a compact skull, a relatively short muzzle and wrinkled skin on the forehead

EYES
Rounded in shape and dark in colour

EARS
Set high towards the rear of the head, rather triangular in shape

CHEST
Relatively broad

TAIL
Set high, thick at its base, and carried away from the body

WEIGHT
Variable, but typically averaging 8–13.5 kg (18–30 lb)

GROOMING

FEEDING

EXERCISE

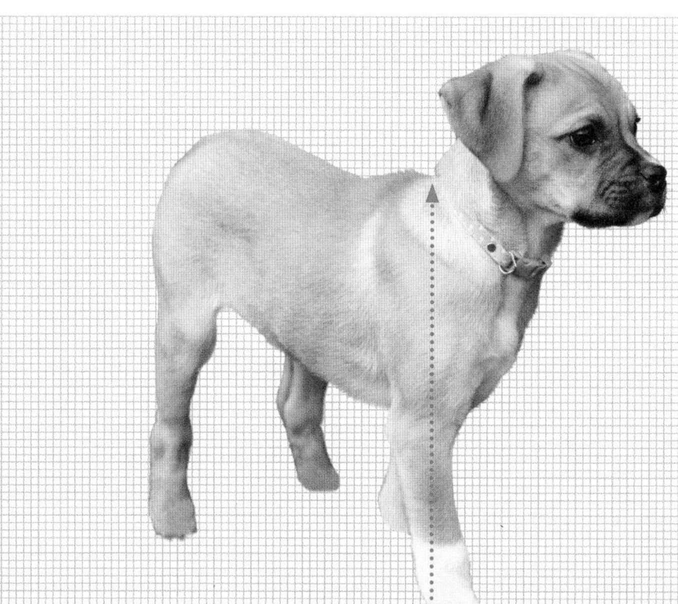

HEIGHT AT SHOULDER
Variable, but usually 33–38cm (13–15 in)

CAVAPOO
Cavalier King Charles Spaniel x Poodle

The first Cavapoos were bred back during the 1950s in the USA, and have since built up a worldwide following. In Australia, they are better-known today as Cavoodles. Their coat type is surprisingly variable, but may well require some professional grooming to keep them looking at their best. Try to obtain a Cavalier King Charles (see pp. 76–7) crossed with a Miniature rather than a Toy Poodle, as they tend to be less vulnerable to health issues. Cavapoos generally learn quickly, but can suffer from boredom on occasions if ignored.

PERSONALITY

Cavapoos are intelligent by nature and learn fast. They display a strong affinity for people, possessing a playful disposition and make an ideal, relatively small-sized companion.

HEALTH AND CARE

Veterinary check-ups are essential, especially since Cavalier King Charles Spaniels are susceptible to a range of health issues. Not all of these, such as mitral value issues affecting the heart, will be discernible immediately in a puppy, so regular checks can be advantageous.

AS AN OWNER

It is important to ensure that a Cavapoo has regular daily exercise, so it does not become obese, which will otherwise exacerbate any predisposition to any mitral value problem. Train your dog to be left for periods – Cavapoos can otherwise easily suffer separation anxiety.

Specification

Rather tousled, with a relatively large head and a short muzzle, which can leave them vulnerable to heat stroke. Their eyes can be very expressive and appealing.

RECOGNITION D
Not a recognized breed at present

LIFESPAN
13–15 years

COLOUR
Variable – often chestnut with white markings, but can be single colours, such as white or even black, reflecting the poodle input into the Cavapoo's ancestry

HEAD
Short, with a relatively short muzzle and wrinkled skin on the forehead

EYES
Deep set and dark in colour

EARS
Set well-back on the head, and hang down

CHEST
Medium width

TAIL
Set high, thick at its base, and can be carried over the back

WEIGHT
Depends on the poodle used. Typically varies from 5.5–11.5 kg (12–25 lb)

CHILD FRIENDLINESS

GROOMING

FEEDING

EXERCISE

HEIGHT AT SHOULDER
Depends on the poodle used. Variable, but usually 25–35.5cm (10–14 in)

POMCHI
Pomeranian x Chihuahua

This cross was first carried out in the USA, and then started to become popular in the UK from the 1980s onwards. Do not be fooled by the small size of Pomichis though – they are actually very assertive by nature, although they also make very loyal companions as well. They can also be noisy, being prone to excessive barking, and this behaviour should be deterred.

PERSONALITY

Playful and intelligent, Pomchis will form a close bond with those in their immediate family, but they are far more likely to be withdrawn towards strangers, and will prove to be alert guardians. When exercising Pomchis, take care to avoid areas where there are large dogs that may chase them.

HEALTH AND CARE

Patellar luxation, affecting the kneecaps, is relatively common in both ancestral breeds, and dental issues are well-documented too, and unsurprisingly, these can crop up in Pomchis as well.

AS AN OWNER

These small dogs are very suitable for a home with older children, but are not so suitable where there are youngsters who may handle them roughly. Pomchis learn fast, but can display a stubborn side to their nature, seeking to get their own way, so be firm and consistent when training these dogs, in the hope of avoiding this issue.

Specification

Much depends as to whether a short- or long-coated Chihuahua contributed to an individual's ancestry. Pomchis can have either single or double-layered coats, although they usually have a profusely furred tail.

RECOGNITION D
Not a recognized breed at present

LIFESPAN
14–17 years

COLOUR
A wide range, reflecting the Chihuahua input into their ancestry. They can be self-coloured, or have white areas in the coat. Sable and merle individuals are also known

HEAD
Rounded between the ears, with the skull being rather domed and narrow muzzle

EYES
Deep set and dark in colour

EARS
Raised vertically and triangular in shape

CHEST
Covered with longer hair than on the body

TAIL
Set high and held upright with a long plume of hair that hangs down over the side of the body

WEIGHT
Typically ranges from 2.5–6.5kg (6–14 lb)

CHILD FRIENDLINESS

GROOMING

FEEDING

EXERCISE

HEIGHT AT SHOULDER
Variable, but usually 15-25cm (6-10 in) 15-25 cm

SCHNOODLE

Schnauzer x Poodle

Often blessed with highly appealing looks, Schnoodles will settle into the home very well, making excellent companions, and this is a common trait that characterizes the most popular designer dogs. Schnoodles are intelligent, lively and responsive dogs by nature, with crosses of this type having originally been made around 40 years ago.

PERSONALITY

These dogs can be quite determined, and also may not agree very well with cats, although a puppy brought up in a home along an established resident feline should not be a problem in this regard.

HEALTH AND CARE

Breeding stock should have been tested for known weaknesses, such as ocular issues like hereditary cataracts. Joint issues too, such as hip dysplasia (for which a puppy's parents should again have been screened) and patellar luxation also need to be borne in mind.

AS AN OWNER

Schnoodles are ideal family pets, being generally affectionate and well-disposed to children. They are invariably keen to play, although play sessions may become rather boisterous on occasions with younger companions, and monitoring may be required!

Specification

It is vital to find out the contributing breeds precisely, as this will give the best clue as to the likely adult size of a puppy. Their appearance is very variable, but their round eyes invariably stand out under their profuse eyebrows.

RECOGNITION D
Not a recognized breed at present

LIFESPAN
10–15 years

COLOUR
Very variable, as in the case of the parent stock. Black and combinations thereof, such as black and tan, are quite often seen, and other possibilities include apricot, silver, grey, sable and more

HEAD
Broad, with a relatively powerful muzzle and a rather domed top to the skull

EYES
Deep set and dark in colour, beneath distant eyebrows

EARS
Broad, rather triangular and hang down the sides of the head

CHEST
Medium width, may be slightly oval in appearance

TAIL
Set high, and held upright

WEIGHT
Exceedingly variable, depending on the exact cross

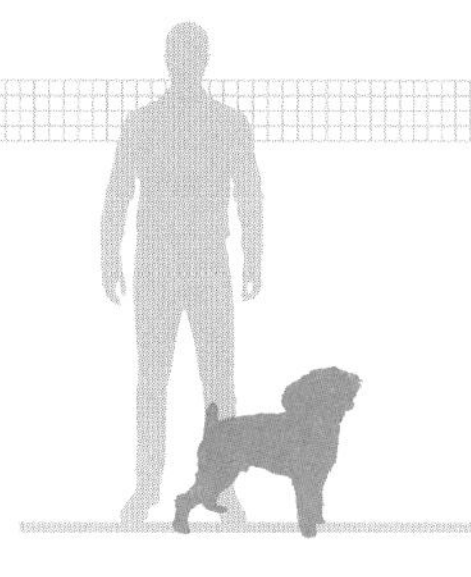

CHILD FRIENDLINESS

GROOMING

FEEDING

EXERCISE

HEIGHT AT SHOULDER
Very variable, depending on breed combination used

RESOURCES

FURTHER READING

The Atlas of Dog Breeds of the World by Bonnie Wilcox & Chris Walkowicz (TFH Publications, 1989).

The Canadian Kennel Club Book of Dogs by The Canadian Kennel Club (Stoddard Publishing, 1988).

Canine Lexicon by Andrew De Prisco & James B. Johnson (TFH Publications, 1993).

The Complete Book of Australian Dogs by Angela Sanderson (The Currawong Press, 1981).

The Complete Dog Book by The American Kennel Club (Ballantine Books, 2006).

The Complete Dog Book by Peter Larkin & Mike Stockman (Lorenz Books, 1997).

Dictionary of Canine Terms by Frank Jackson (Crowood Press, 1995).

The Dog: The Complete Guide to Dogs & Their World by David Alderton (Macdonald, 1984).

Dogs: The Ultimate Dictionary of Over 1000 Dog Breeds by Desmond Morris (Ebury Press, 2001).

Gun Dog Breeds. A Guide to Spaniels, Retrievers and Sporting Dogs by Charles Fergus (Lyons & Burford, 1992).

Herding Dogs: Their Origins and Development in Britain by I. Combe (Faber & Faber, 1987).

Hounds of the World by David Alderton (Swan Hill Press, 2000).

The Kennel Club's Illustrated Breed Standards by The Kennel Club (Ebury Press, 1998).

Legacy of the Dog: The Ultimate Illustrated Guide to over 200 Breeds by Testsu Yamazaki (Chronicle Books, 1995).

The New Terrier Handbook by Kerry Kern (Barron's, 1988).

Smithsonian Handbooks: Dogs by David Alderton (Dorling Kindersley, 2002).

Toy Dogs by Harry Glover (David & Charles, 1977).

KENNEL CLUBS

American Canine Hybrid Club,
10509 S & G Circle,
Harvey, AR 72841, USA.
www.achclub.com

American Kennel Club, 260 Madison Avenue,
New York, NY 10016, USA.
www.akc.org

Australian National Kennel Council, PO Box 285,
Red Hill South, Victoria 3937, Australia.
www.ankc.org.au

Canadian Kennel Club, 89 Skyway Avenue, Suite
100, Etobicoke, Ontario M9W 6R4, Canada.
www.ckc.ca

Continental Kennel Club, PO Box 1628, Walker,
LA 70785, USA.
www.ckcusa.com

Irish Kennel Club, Fottrell House, Harold's Cross
Bridge, Dublin 6W, Republic of Ireland.
www.ikc.ie

The Kennel Club, 10 Clarges Street, London
W1J 8DH, England.
www.thekennelclub.org.uk

National Kennel Club, 255 Indian Ridge Road,
PO Box 331, Blaine, Tennessee 37709, USA.
www.nationalkennelclub.com

Dogs NZ, Prosser Street,
Private Bag 50903, Porirua 6220, New Zealand.
www.nzkc.org.nz

United Kennel Club, 100 East Kilgore Road,
Kalamazoo, MI 49002, USA.
www.ukcdogs.com

Universal Kennel Club International,
101 W Washington Avenue, Pearl River,
NY 10965, USA.
www.universalkennel.com

World Wide Kennel Club, PO Box 62, Mount
Vernon, NY 10552, USA.
www.worldwidekennel.qpg.com

INDEX

ACKNOWLEDGMENTS

With thanks to the owners for permission to photograph the following:

p. 1 CH Gameaway Kiwi JW; p. 2–3 (left to right) Karak Lord of The Isles, Roybridge Ruff Diamond JW ShCM, Helmlake Titan, ShCH Lokmadi Nick The Brief, Inzievar Silver Gilt, Bonvivant Moonrush; p. 5l CH Lunabrook Hot Shot ShCM; p. 5r Ringlands Golden Dollar; p. 10r Embeau Rive Gauche At Jethard; p. 11 ShCH Jennaline Ello Ello Ello; p.12l Bonvivant Moonrush; p. 12m Embeau Rive Gauche At Jethard; p. 12r CH Vanitonia Well Did You Evah; p. 13tr ShCH Lokmadi Nick The Brief; p. 13br Helmlake Titan; p. 13 bm ShCH Goldenquest Ambassador JW; p. 13bl Moonreapers Vandal; p. 18–19 (left to right) Helmlake Titan, Karak Lord of The Isles, ShCH Lokmadi Nick The Brief, Inzievar Silver Gilt, Bonvivant Moonrush; p. 20–21 Natimuk Keenaughts Bobby Dazzler; p. 22–23 Beauview Hot Gossip; p. 24–25 CH Drakesleat Otto Bahn; p. 26–27 Marshdae Bobbi-Bobbi With Mazeena; 28–29 Inglecourt Magical Annie; 30–31 Lireva's The Future's In Focus; 32–33 Benatone Love Me Tender Abbyat; 34 Benatone Love Me Tender Abbyat; 36–37 Chelanis In Cahoots; 38–39 CH Warmingham Looks To Thrill At Ashoka; p. 40–41 CH Avonbourne Mandzari At Chobrnag JW ShCM; p. 42–43 Fourheatons Grandopera Libertine; p. 44–45 Sophyla Sable; p. 46–47 Kensing Henry Higgins JW ShCM; p. 48–49 Jaeva Jingle Bell Rock With Zippor; p. 50–51 CH Gameaway Kiwi JW; p. 52–53 Berrybrezze Illumination; p. 54–55 Jenina Jay Jay; p. 56–57 CH Krisma Jammy Dodger; p. 58–59 Cotonneux Athos; p. 60–61 CH Vanitonia Well Did You Evah; p. 62 CH Bymil Sherry Twist; p. 64–65 Lokmadi Kabuki JW ShCM; p. 66–67 Bakalo Shayna Punim At Shanita; p. 68–69 CH Delwins Paddy O'Reilly; p. 70–71 Ringlands Golden Dollar; p. 72–73 CH Quiche's Douglas At Kanix IMP CAN; p. 74–75 Vanitonia Blue Murder; p. 76–77 CH Miletree Nijinski JW; p. 78–79 CH Mid Shipman Of Vardene; p. 82–83 CH Nobozz Drunk and Disorderley; p. 84–85 Inzievar Silver Gilt; p. 86–87 Moonreapers Vandal; p. 88–89 Trigger At Sunrise; p. 90–91 Sangria Fabiola; p. 92–93 Peerieglen Pearl JW; p. 94–95 Kokonor Always Ina Pickle To Ludgate; p. 96–97 CH Nocte Boris Karloff; p. 98–99 Switherland Double Delight At Feorlig; p. 102–103 Araidh Ditto; p. 104–105 Ashleyheath Pot of Gold; p. 106–107 CH Bellpins Wear The Fox Hat; p. 108–109 Belfox Belstar; p. 110–111 CH Saradon Dressed To Impress; p. 112–113 Wynele Mz Dee Meana; p. 114 Naka No Benihana Go Kazusa; p. 116–117 Falconers Cotillon Of Ware; p. 118–119 Embeau Rive Gauche At Jethard; p. 120–121 Rossvale Latin Lover; p. 122–123 Linsdown Georgia With Strco; p. 124 CH Snowmeadow Jellybean Jilly JW ShCH; p. 126–127 Morrow Blue Eucalyptus At Gilsand; p. 128–129 CH Highclare Energizer; p. 130–131 CH Forstal's Kaliznik; p. 132–133 ShCH Brittyhill Sorrel; p. 136–137 Mabina Made At Inzadi; p. 138–139 Lotsmoor Wild Passion; p. 140–141 CH Gameaway Kiwi JW; p. 144–145 Chopan Miss Chandelle; p. 148–149 Benninghof Cody; p. 150–151 CH Moonhill's Forever Classic; p. 152–153 Elaridge Quincie JW; 154–155 CH Roybridge Ruff Diamond JW ShCM; p. 156–157 NED/BELG CH Ravenstone Bersin Krisen; p. 158–159 CH Lunabrook Hot Shot ShCM; p. 160–161 Arnac Bay Pride; p. 162–163 Beauvallon Philanderer JW; p. 164–165 ENG & AM CH Joval Jumpin Jack Flash At Jokyl; p. 166–167 ShCH Jennaline Ello Ello Ello; p. 168–169 Nicael Oak At Helmlake; 170–171 ShCH Goldenquest Ambassador JW; 172–173 CH Carpenny Anchorman; p. 174–175 CH Tartarian Gold Dust; p. 176–177 CH/FCI INT/NED/BELG CH Fullani Flekkefjord; p. 178–179 CH Bareve Bandari JW; p. 180–181 Branchalwood Aylanmhor For Burpham JW ShCM; p. 182–183 Bournhouse Paper Parade For Oaktrish; p. 184–185 Blondie's Read My Lips For Bymil; p. 186–187 Darahill Hufflepuff At Sablemyst; p. 188 Stormdancer Apple Turnover To To To Castanho; p. 190–191 Schutzer Valentino; p. 192–193 ShCH Abelard Scotch Poacher By Brobruick; p. 194–195 Chayo Blue Thunder; p. 196–197 Chigwri Uncanny; p. 198–199 Pines Mercedes; p. 200–201 Chiorny Lider Of Potterspride; p. 202–203 ShCH Lokmadi Nick The Brief; p. 206–207 29 ShCH Tefryn Mackinley At Beltane; p. 208–209 CH Mirengo's Muko Mako; 210–211 CH Coxellot Buttons And Beaux By Flintstock; p. 212–213 Karak Lord of The Isles; p. 216–217 CH Redwitch Heaven Can Wait; p. 218–219 Rayvonley Lion; 220–221 ShCH Carnbargus Congratulation JW; p. 222–223 IR CH Darmarnor Bertie Bear JW ShCM; p. 224–225 CH Poolsway I've Got Spirit; p. 226–227 CH Mountcook Tiger Lily IMP; p. 228–229 Strelkos Sillmarion; p.230–231 Osemans Percyverance At Sargebilko; p. 232–233 CH Boughton Benvoluto; p. 234–235 Cenninpedr Ferch Bert At Fearnaught; p. 236–237 Helmlake Titan; p. 238–239 Borka Sequel

Key to abbreviations *CH= UK Champion, otherwise prefixed by the country, e.g., AM CH=American Champion; ShCH=Show Champion; ShCM=Show Certificate of Merit; JW=Junior Warrant; IMP=Imported, often followed by country, e.g., AM=USA, BELG=Belgium, CAN=Canada, FR=France, NED=Netherlands*

Pictures *The publishers would like to thank the following for permission to use their images:*

p. 4 Masarik/Shutterstock; p. 6–7 Chris Collins; p. 8 Getty Images/ Sharon Montrose/The Image Bank; p. 9tl Phil Banko; p. 9tr GK Hart/ Vikki Hart/The Image Bank; p. 9m LWA-Sharie Kennedy/zefa; p. 9b Herbert Spichtinger/zefa; p. 10l mariait/Shutterstock; p. 14 Farlap/ Alamy; p. 15tl Barrie Harwood/Alamy; p. 15tr Flyby Photography-www.flybyphotography.co.uk/Alamy; p. 15b Gem Russan/Alamy; p. 35 Konstantin Gushcha/Shutterstock; p. 63 SkaLd/Shutterstock; p. 80–81 cynoclub/Shutterstock; p. 100–101 Eudyptula/Shutterstock; p. 115 eAlisa/Shutterstock; p. 125 dien/Shutterstock; p. 134 Jne Valokuvaus/Shutterstock; p. 135 Tikhomirov Sergey/Shutterstock; p. 142–143 Erik Lam/Shutterstock; p. 146–147 Dora Zett; p. 189 Kuznetsov Alexey; p. 204–205 Dora Zett/Shutterstock; 214–215 cynoclub/Shutterstock; p. 240 SikorskiFotografie/Getty; p. 241 Bob and Pam Langrish KA9Photo; p. 242–243 Nynke van Holten/ Shutterstock; p. 244 Imagebroker/Alamy; p. 245 Graham Johns/ Alamy; p. 246–247 Bob and Pam Langrish KA9Photo/Alamy; p. 248–249 Flabygasted/Shutterstock; p. 250 CLS Digital Arts/ Shutterstock; p. 251 Danita Delimont/Alamy